ANDOVER COLLEGE
065697

D1389460

THE GENIUS TEST

THE GENIUS TEST

WITHDRAWN

GILES SPARROW

Quercus

Contents

POLITICS AND ECONOMICS

MATHEMATICS

PHYSICS

COSMOLOGY

Are you a genius?

Check off the topics as you master them in increasing level of difficulty.

Mind-blowing

- ☐ I think, therefore I am
- ☐ Free will and God
- ☐ The hard problem
- ☐ Structuralism and semiotics
- ☐ Fermat's Last Theorem
- ☐ Gödel's incompleteness theorems
- ☐ Theories of everything

Formidable

- ☐ Human genome
- ☐ The search for understanding
- ☐ Good and evil
- ☐ Existentialism
- ☐ Artificial intelligence
- ☐ Postmodernism
- ☐ Globalization and nationalism
- ☐ Keynesianism and monetarism
- ☐ Postcapitalism
- ☐ The Riemann hypothesis and Goldbach's conjectures
- ☐ Schrödinger's cat
- ☐ The Higgs boson
- ☐ Special and general relativity
- ☐ Black holes
- ☐ Multiverses

Tough

- ☐ Evolution
- ☐ Genes and DNA
- ☐ Language and consciousness
- ☐ Genetic engineering
- ☐ The nature of reality
- ☐ Modern art
- ☐ Literary criticism
- ☐ Modern architecture
- ☐ Democracy
- ☐ Conservatism, liberalism and socialism
- ☐ Macro- and microeconomics
- ☐ Environment and climate change
- ☐ Infinity
- ☐ Probability and statistics
- ☐ Quantum physics

Tricky

- ☐ Origins of life
- ☐ Human origins
- ☐ Nature vs nurture
- ☐ The human brain
- ☐ Psychology
- ☐ History of art
- ☐ History of literature
- ☐ Digital politics
- ☐ Capitalism
- ☐ Chaos
- ☐ Nanotechnology
- ☐ The Big Bang
- ☐ Life in the Universe

Introduction

‘So, what do *you* think about the Higgs boson?’

We've all had that experience – nodding along wisely to a conversation about a topic that we barely understand, when someone asks for our opinion, and the floor abruptly drops away from under us. Usually, our instinct is to mutter something noncommittal or agree with whoever seems to be the smartest person in the group. But what if we could *be* that person? The one with an informed opinion on everything, from the Higgs boson and the crisis of capitalism, to genetic engineering and postmodernism?

That might sound like a daunting task – the modern world is an extraordinarily complex place and who among us really has the time or energy to devote to really understanding its complexities, let alone the deep history of ideas that underlie and shape today's society?

So we came up with *The Genius Test* – a mental gymnasium that will help you master the essentials of a huge variety of topics ranging across fundamental concepts in science, philosophy, the arts and politics (on the opposite page you can see how we've rated them, from the merely tricky to the truly mind-blowing).

In each chapter, five questions ask 'Are you a Genius?' helping gauge your understanding of the subject before plunging in, or check you've mastered it afterwards. (Answers are on the last page of each chapter, and you may even pick up some interesting additional facts along the way.) 'Ten Things a Genius Knows' offer a thorough overview of the topic, helping you get to grips with its central ideas and historical development in no time at all. 'Talk like a genius' provides you with handy conversational snippets – opinions, facts and intriguing asides to burnish your credentials as the smartest person in the room. There's even a handy 'Bluffer's summary' boiling the whole topic down to a couple of sentences that could help get you out of embarrassing situations.

This book won't, on its own, make you a bona fide genius of course, but it's a good start. It'll make you a better bluffer at parties, and maybe it'll even change the way you look at the world and reveal a previously undiscovered intellectual ability or interest. As lexicographer Samuel Johnson put it, 'the true genius is a mind of large general powers, accidentally determined to some particular direction.' Who knows in what direction *The Genius Test* will take you?

Origins of life

'The cosmos is within us. We are made of star-stuff. We are a way for the Universe to know itself.'

CARL SAGAN

Our planet coalesced from a cloud of dust and gas orbiting the young Sun about 4.5 billion years ago. Conditions on its newly formed surface would have been incredibly hostile, yet life seems to have taken hold surprisingly quickly (even if it took a very long time to develop beyond single-celled organisms). So how exactly did life begin? It's a mystery that has puzzled leading scientists for generations and has given rise to some extraordinary ideas.

Where did life begin – on the shorelines of the first seas, in the cold depths of primeval oceans or, perhaps, on another planet entirely?

1 A laboratory experiment carried out in the 1950s produced the essential components of DNA.

TRUE / FALSE

2 Biologists believe that our earliest ancestors might be related to microbes called Archaea, that today live only in environments of extreme heat or acidity.

TRUE / FALSE

3 Carbon and water are absolutely necessary building blocks for life – without them, complex biochemistry is impossible.

TRUE / FALSE

4 Single-celled organisms have the potential to survive the journey between planets, perhaps seeding the Earth with life.

TRUE / FALSE

5 Complex multicellular life took hold on Earth in an event called the 'Cambrian explosion' about 540 million years ago.

TRUE / FALSE

TEN THINGS A GENIUS KNOWS

1 What is life?

Although opinions differ wildly on specific details, most biologists would probably agree with the broad definition of a living organism as a self-organizing system that can harvest energy from its environment in order to sustain itself, grow and reproduce, and adapt to that environment. In practice, this means harnessing a variety of complex chemical reactions within a hospitable environment known as a 'cell'. Cells come in a variety of more or less complex forms and are the essential building blocks of life, so when we ask how life got started, we're really investigating the origin of the first cells.

2 Earliest evidence

Dating of ancient rocks from Earth and meteorites (space rocks that have remained unaltered from the earliest times) suggests that our planet formed about 4.6 billion years ago, initially with a molten surface. Heavy bombardment by large asteroids continued until at least 3.8 billion years ago, but the earliest fossilized life, the remnants of microbe colonies called stromatolites are found just a few hundred million years later, in rocks about 3.5 billion years old. What's more, in 2015 geochemists found traces of chemicals that appear to have been made by living organisms sealed inside 4.1-billion-year-old zirconium crystals. So how did life get a foothold so fast?

3 Primordial soup

In a letter of 1871 (a dozen years after publishing his theory of evolution by natural selection), Charles Darwin speculated that life could have got started in a 'warm little pond' on the still-cooling surface of the ancient Earth. This theory caught the imagination of many scientists, and is generally called the 'primordial soup' hypothesis (though that term was not coined until the 1920s). Water is certainly a must-have for life – complex chemicals have no hope of forming unless their building blocks can move around in some kind of solution, to encounter and react with each other. Fortunately, water is one of the best solvents there is, and Earth has no shortage of it.

4 The Miller-Urey experiment

In 1952, US biochemists Stanley Miller and Harold Urey famously attempted to recreate conditions in the primordial soup by passing steam through a mix of hydrogen, methane and ammonia gases (thought to be likely components of Earth's early atmosphere) and energizing the whole mixture with occasional electrical sparks of faux-lightning. After a week, the condensed liquid was analysed – Miller reported finding at least three amino acid molecules (vital building blocks on the road to life) and possibly a couple more. After Miller's death, in 2007, scientists reanalysed samples sealed since the original experiment using more sensitive techniques, and found no fewer than 20 different amino acids.

5 Building life

Many chemists have followed in the footsteps of Miller and Urey with more sophisticated experiments designed to better mimic our improved understanding of Earth's early environment. It seems beyond doubt that fairly simple chemical reactions could have produced a soup of simple carbon-based 'organic' molecules in short order (carbon is vital to life because it forms the widest variety of chemical bonds of any common element). The big challenge, however, is getting from those simple building blocks to complex self-replicating molecules such as DNA (see page 17). The establishment of molecules whose reproduction could be shaped by 'selection pressures' would be a big step on the road to life, but some scientists question whether random chemical encounters in the primordial soup could have reached this level of complexity in the relatively narrow window of time between the Earth's formation and the first fossil evidence.

6 Black smokers

One popular solution to this problem shifts the birth of life from shallow surface waters to deep oceans, where volcanic vents called 'black smokers' spew a rich mix of chemical nutrients into the cold, dark waters. Discovered in the 1970s, smokers are stalagmite-like mineral pillars that play host to entire ecosystems that thrive without the need for warmth or light from the Sun. In recent years, some biologists have speculated that microscopic pores within the smokers could have acted as natural cells for incubating the first stages of life, trapping a rich stew of chemicals – including organics drifting down from the surface above – in an energy-rich environment ideal for the fast-track development of complex chemistry.

7 Pangenesis

Another possible way around the problem of life's rapid appearance is to assume that it didn't start on Earth at all. The 'pangenesis' hypothesis suggests instead that the building blocks of life are scattered throughout our galaxy, and that the meteorites and comets that bombarded our newborn planet also supplied a ready-made starter kit of organic chemicals, and perhaps even entire deep-frozen cells. Supporters of pangenesis argue that it provides several billion more years for random chemical reactions to chance upon the formula for life. While this might sound far-fetched, astronomers ***have*** identified increasingly complex organic molecules in comets and interstellar dust clouds. What's more, large meteorite impacts are now known occasionally to transfer rocks between the planets of our solar system, and there's evidence that some Earth microbes and even more complex forms of life can survive the hostile conditions of interplanetary space for surprisingly long periods of time.

8 The earliest organisms

The first forms of life are thought to have fallen in two broad 'domains': the Archaea and the Eubacteria. Both groups were single-celled organisms, though some clustered together to form larger colonies. Archaea use a wide variety of different chemical and metabolic pathways to 'make a living' from their environment. Today, they are found in a huge range of conditions including some, such as hot acid springs and black smokers, that were once thought inimical to life. Eubacteria, in contrast, have more familiar metabolic processes including respiration, photosynthesis and fermentation. They are found in a more limited range of 'hospitable' environments, yet curiously, genetic evidence suggests that our own domain, the complex organisms known as 'Eukarya', are actually more closely related to the Archaea than the Eubacteria.

9 The oxygen catastrophe

Early conditions on Earth were very different to those that dominate today. There was very little free oxygen in the atmosphere. From at least three billion years ago, photosynthetic Archaea and Eubacteria thrived through photosynthesis, absorbing carbon dioxide and pumping out oxygen. About 2.3 billion years ago, however, oxygen levels in the atmosphere soared, turning the air toxic for many early life forms and paving the way for a new metabolic pathway that uses oxygen to release energy from chemicals and is used by today's animals: respiration.

10 Endosymbiosis and more complex life

Our own domain of life, the Eukarya, is distinguished by a much more complex cell structure, including the presence of a nucleus that holds most of the cell's genetic information. Most biologists believe that the first 'eukaryotic' cells arose as specialized microbes absorbed each other with mutually beneficial results – a sequence of events called 'endosymbiosis'. All large multicellular life forms are eukaryotes with a common ancestor going back 1.6 to 2.1 billion years. However, they remained largely single-celled until about 575 million years ago, when the first fossils of larger and more complex life, the curious pillow-like creatures known as the 'Ediacaran biota', appear.

TALK LIKE A GENIUS

‘ One of the most amazing things about life on Earth is that it only took hold once – go back far enough and genetic mapping shows that everything’s descended from a single common ancestor, probably a simple bacterium. Why should life get started just once in those first few hundred million years, and never again? The answer’s probably that the descendants of that first bacterium made it impossible for any other attempts to get a foothold – not so much survival of the fittest, as survival of the first. ’

‘ The big chicken-and-egg problem is that you need DNA in order to make proteins, but you also need the right proteins to replicate DNA. The big challenge for biologists is coming up with a pathway to make something like simple DNA or proteins without life already existing. ’

‘ The panspermia theory might sound outlandish, but don’t forget we’re finding more and more environments in the solar system that might be suitable for life. ’

WERE YOU A GENIUS?

1 FALSE – the Miller-Urey experiment made amino acids, but those aren’t components of DNA. In 1961, however, Spanish scientists did make DNA components in a similar experiment.

2 FALSE – it’s true that Archaea are closer to us than Eubacteria, but they’re actually quite widespread in a whole range of environments.

3 TRUE (probably) – carbon is a must-have to form complex compounds, but liquids other than water can act as chemical solvents in cold environments.

4 TRUE – some microbes can survive exposure to space, although we don’t know if they’d last for the millions of years it might take to drift between planetary orbits.

5 TRUE – although multicellular life had begun to develop tens of millions of years earlier, the Cambrian explosion is a hugely important event that gave rise to most modern animal groups.

It’s easy to make simple organic chemicals, but bridging the gap to the complex biochemistry of life is a huge leap.

Evolution

'The nature of the Universe loves nothing so much as to change the things which are, and to make new things like them.'

MARCUS AURELIUS

Charles Darwin's theory of evolution by natural selection may be the greatest single achievement in the history of science, a simple and elegant model that explains the huge variety of life on Earth. More than 150 years from its initial publication, however, it remains controversial in some quarters because it threatens to undermine religious views of creation. And evolutionary biologists are still ironing out the wrinkles as to how it all actually works in practice.

Darwin's basic idea is easy to grasp, yet hugely powerful when it comes to explaining the natural world – can you get to grips with its implications?

1 In order to take place, evolution requires the remixing of genetic information that happens during sexual reproduction.

TRUE / FALSE

2 Evolution will always favour traits that help an individual to survive and reproduce, and 'weed out' characteristics that harm the chances of reproduction over time.

TRUE / FALSE

3 Scientists today try to use the principles of evolution as the starting point for their classification of different species into larger groups.

TRUE / FALSE

4 Darwin realized that the different bills of various finches on the Galápagos Islands were a sign that they had evolved from a common ancestor that colonized the islands from the South American mainland, but he didn't offer an explanation for what had driven their evolution.

TRUE / FALSE

5 Sexual selection can sometimes lead to features that actually hamper the day-to-day survival ability of some animals.

TRUE / FALSE

TEN THINGS A GENIUS KNOWS

1 How evolution works

The theory of evolution by natural selection argues that the characteristics of organisms change from generation to generation as a result of 'selection pressures'. These range from environmental conditions and predator abilities to the sexual preference of potential mates, and operate on the random differences that arise from time to time out of genetic mutation and the mixing of characteristics from parent organisms. If an individual's particular set of characteristics make it better suited than others of its species to a particular time and location, then it's more likely to survive, breed and pass those characteristics on to its offspring.

2 Speciation

Genetic studies show that all life is ultimately descended from a single common ancestor, so how did we arrive at the huge variety of life we know today? The answer is that as life spread, different selection pressures operated on different organisms. Sometimes, new pressures arise due to changes in the surrounding environment, while at other times a chance mutation offers an organism a new way to make a living that its offspring can further exploit (perhaps taking them out of direct competition for resources with their cousins). Over many generations, different pressures acting on different organisms can give rise to populations whose genes are so different from each other that they can no longer produce viable offspring together – the textbook definition of separate species.

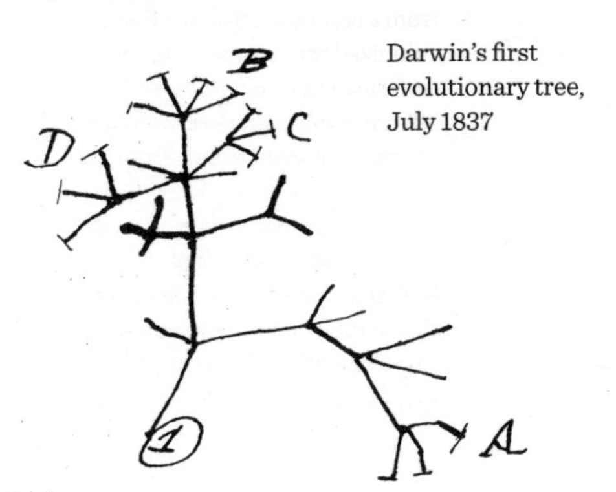

Darwin's first evolutionary tree, July 1837

3 Early theories of life

From about the 15th century onwards, voyages of discovery, trade and colonization made European scholars aware, for the first time, of the huge variety of life on Earth. In 1735, Sweden's Carolus Linnaeus invented the binomial classification of species, grouping organisms into genera, families, classes and orders at increasingly higher and more inclusive levels. The resulting 'tree of life' looks irresistibly like a genealogical table, raising obvious questions as to whether similar modern species might be diverse descendants from a single ancestor. Around the same time, the growth of large-scale mining during the Industrial Revolution led to an improved understanding of the rocks beneath our feet, revealing both the Earth's great age, and fossils from extinct species. This discovery of 'Deep Time' provided an enormous span for evolution to do its work.

4 Darwin's story

A naturalist since childhood, Charles Darwin gathered much of the evidence for his theory during the second South American survey expedition of HMS *Beagle* (1831–36). He collected huge numbers of specimens, unearthed countless fossils, and visited the Galápagos archipelago, where he famously observed the unique variations between species of finch and giant tortoises on different islands. Back home in England, he researched the selective breeding of domesticated animals and read economist Thomas Malthus's alarming theories on the dangers of runaway human population growth. Considering the issue of competition for limited resources, Darwin formulated his key idea of selection pressures allowing species to diversify over time and thus avoid direct competition. He finally published his ideas in *On the Origin of Species* (1859).

5 Reception to Darwin

Darwin's idea shocked Victorian society to the core – and there are plenty who still dislike its implications. He initially dodged the issue of human origins, but later confronted it head-on in *The Descent of Man* (1871). More than a century earlier, Linnaeus had grouped humans among the primates, alongside monkeys and great apes, and Darwin's work concluded that we shared common ancestors with our primate cousins at various points in the distant past. Despite the widespread parodies of the time, Darwin did not assert that we were 'descended' from apes – rather that we are their distant cousins.

6 Neodarwinism and the selfish gene

Darwin had little to say on the mechanism by which adaptations were passed from generation to generation. It was only in the early 20th century that the theory was placed on a firmer footing with the widespread recognition of genes (see page 17). Concepts of genetic mutation and diversity within populations allowed biologists such as J.B.S. Haldane to build mathematical models of evolution, an approach called 'Neodarwinism'. Various speciation mechanisms were discovered, but problems remained – particularly where an organism's adaptations seem to make its individual survival *less* likely. From the 1960s this led to the 'selfish gene' hypothesis – the idea that evolution maximizes the spread of useful genes through a population, rather than necessarily benefiting individuals.

7 The pace of evolution

The simplest interpretation of Darwinian evolution, known as 'phyletic gradualism', involves change at a slow but steady pace. Species gradually transform over time, and new ones appear, as selection pressures act on genetic mutations that pop up more or less at random. At the other extreme is 'punctuated equilibrium' – a model in which species remain essentially the same over long periods of time, before going through periods of rapid change and diversification in response to environmental crises. The discovery that catastrophic events, such as meteor impacts and changes in climate, coincide with major changes in the fossil record certainly suggests that punctuated equilibrium has a role to play.

8 Missing links aren't missing

Some religious creationists take a simplistic view that God created our ecosystem wholesale, including fossils and genetic templates to test our faith. More often, however, opponents of Darwinian evolution like to raise what feel like 'scientific' problems – such as an apparent lack of evidence showing important changes in the evolutionary record ('missing links'). Such arguments tend to ignore fossil finds that do indeed go some way to bridging these gaps, and misunderstand just how fragmented the fossil record is – only a tiny fraction of all species have left fossils behind, and even these are naturally biased towards organisms dying in certain environments. What's more, the 'missing link' complaint is usually based on a gradualist interpretation of evolution that may not be accurate.

9 Why intelligent design isn't true

Another creationist argument, broadly known as 'Intelligent Design' (ID), is that some complex anatomical features could not possibly have sprung into existence through a random mutation offering a selection advantage in a single generation. Instead, there must be some external influence directing the 'shape' of evolution. Common examples of this 'irreducible complexity' include eyes and flight feathers. However, ID advocates often overlook the possibility that such features could have provided different evolutionary advantages in the early stages of their evolution (for example, primitive feathers probably provided small dinosaurs with insulation). What's more, mutations that do not affect an organism's chances of reproducing can remain unnoticed by selection pressures over many generations.

10 Abuses of Darwinism

Darwin's theory is powerful and undeniably true, but all too often, it has been misappropriated for political and ideological reasons. One particular reading depicts evolution as a 'march of progress', with later-evolved organisms necessarily 'superior' to previous ones. This is not the case – organisms are simply adapted to their particular evolutionary niches. Widely used to support assumptions of white European superiority during the 19th-century 'age of empires', this idea lingers on in deep-seated racial assumptions. Meanwhile, proponents of free-market and competition-based economic and social policy often co-opt Darwinian ideas (most notably the phrase 'survival of the fittest'), the scope of which should properly be limited only to biological science.

TALK LIKE A GENIUS

‘A lot of people call it the Darwin–Wallace theory. Darwin knew his theory was dynamite, so he spent two decades refining it and gathering evidence before he was ready to go public. In the end he was nearly beaten to the punch by a naturalist called Alfred Russel Wallace, who had come to the same conclusions while exploring the Malay Archipelago, and actually sent a letter to Darwin for his opinion. Darwin's friends in the scientific establishment engineered a compromise that got both versions out at the same time, establishing that Darwin had got there first. They were lucky that Wallace went along with it when he found out after the fact!’

‘When people say evolution's ‘just a theory’, they're misunderstanding just what a theory is. For scientists, it's a complete description of the way something works, backed up by lots of evidence – you might *refine* it, but you're very, very unlikely to ditch it completely. The proper word for an uncertain, tentative explanation for a set of facts is ‘hypothesis’.’

‘Evolution's often misrepresented in diagrams that show a tree of life with humans at the top, but Darwin had the right idea from the outset. There's a beautiful, simple little diagram in his notebooks from 1837, and it pretty unmistakably shows the pattern of evolution as more of a scrubby, branching bush than a tree.’

WERE YOU A GENIUS?

1 FALSE – evolution can also act on organisms that reproduce asexually, because mutations can take place for other reasons.

2 FALSE – according to the ‘selfish gene’ theory, self-sacrificing traits can arise in which some individuals are lost, while close relatives benefit and pass on the same genes.

3 TRUE – ‘cladistic’ classification attempts to work out how closely species are related by cataloguing their genetic or anatomical features.

4 TRUE – ornithologist David Lack was the first person to identify the bills of Darwin's finches as adaptations to feeding on different plants.

5 TRUE – for example, male peacocks with large tails win the competition to breed despite their tails making them more vulnerable to predators.

When selection pressures affect breeding in one generation, they change the features that the next one inherits.

Genes and DNA

'Rather than believe that Watson and Crick made the DNA structure, I would rather stress that the structure made Watson and Crick.'

FRANCIS CRICK

Deoxyribonucleic acid (DNA for short) is the complex molecule at the heart of genetics – a self-replicating template that carries instructions to make every part of an organism from its constituent protein molecules up to its large-scale anatomy. Although its structure and role in inheritance were established in the 1950s, the technology to map the entire DNA sequence of an organism is much more recent and biologists are still getting to grips with the new understanding unlocked by maps of the entire genome.

If you really want to understand how genetics works, you've got to get to grips with the molecule at the heart of it all.

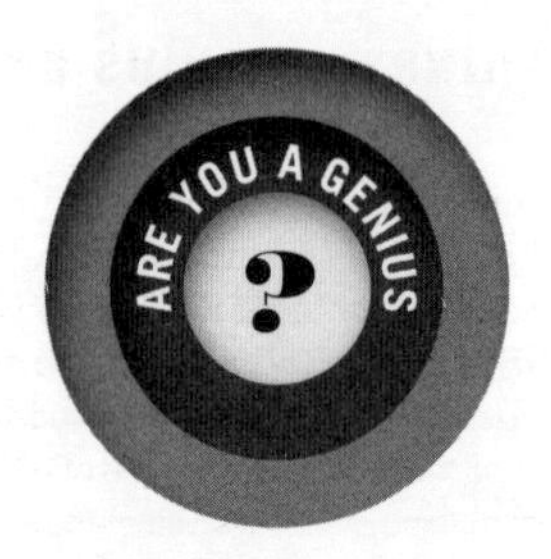

1 Scientists owe much of what they know about DNA function to a small insect called *Drosophila melanogaster*.

TRUE / FALSE

2 During fertilization, two sets of genes from each parent are separated and randomly mixed together in order to determine the single set of genes that will determine the characteristics of offspring.

TRUE / FALSE

3 Each rung on the DNA ladder is made from a pair of amino acids. Some of these can be manufactured in the body but others cannot, so we must get them from our diet.

TRUE / FALSE

4 Like a well-built computer program, DNA has built-in redundancy- and error-checking procedures.

TRUE / FALSE

5 DNA never leaves the nucleus of a cell except when a cell is splitting in two to duplicate itself.

TRUE / FALSE

TEN THINGS A GENIUS KNOWS

1 Genes and chromosomes

There's a lot of jargon associated with genetics, so let's clear up some terms first: a 'gene' is a block of genetic information determining a particular characteristic, such as eye colour, encoded as a sequence of 'base pairs' on a strand of DNA. A 'chromosome', meanwhile, is a single strand of DNA, typically many millions of base pairs long and containing many thousands of genes. The central nucleus in a typical human cell stores 38 chromosomes – 18 identical pairs (for reasons we'll get to later) and two distinct sex chromosomes. The full set of chromosomes is known as the 'karyotype', while the complete set of heritable genetic information contained therein is known as the 'genome'.

2 Discovery of genetics

DNA was discovered in the nucleus of cells by Friedrich Miescher as early as 1869, but its importance was not recognized until much later. Around 1866, meanwhile, Austrian monk Gregor Mendel discovered the principles of genetics through experiments with pea plants. He showed that certain characteristics of the plants were encoded as factors that we now call genes. Each plant carries two versions of every gene (one inherited from each parent), but displays characteristics determined by either one gene or a mix of both. Overlooked at the time, Mendel's work was rediscovered in 1900, when it helped to cement Darwin's theory of evolution. But it wasn't until the 1940s that a trio of American researchers found evidence that DNA molecules were the carriers for genetic code.

3 Genotype and phenotype

An organism's genetic make-up is known as its 'genotype', while its external appearance is its 'phenotype' (some scientists extend the phenotype concept further to include the organism's interactions with its environment). The genotype includes two versions of each gene, but usually only one of these is visible or 'expressed' in its phenotype. The different versions of a gene are known as its 'alleles', and are generally 'dominant' or 'recessive', indicating which will be expressed if both are present. When an organism inherits two alleles of equal rank, the result may be either incomplete dominance (a blend of the two characteristics), or codominance (an expression of both characteristics in different parts of the body).

4 Structure of DNA

The DNA molecule is technically a 'polymer' – a complex molecule made up from a huge number of simple repeating units. Its famous double-helix structure, worked out by Francis Crick and James Watson in 1953 using experimental data gathered by Maurice Wilkins and Rosalind Franklin, is often compared to a twisted ladder. The sides of the ladder are formed from repeating groups known as the 'phosphate backbone', while each rung is made from a pair of chemical bases that join in the middle. There are four possible base units – adenine (A), cytosine (C), guanine (G) and thymine (T), and they bond in specific pairs – adenine and thymine go together, as do cytosine and guanine.

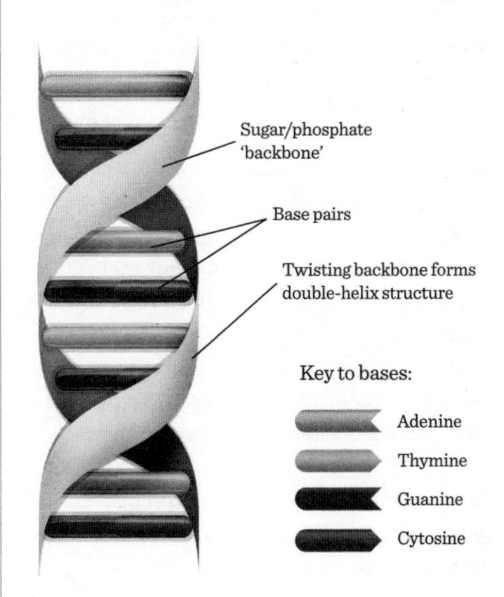

5 The information molecule

Each position on one side of the DNA ladder can be filled with any one of the base molecules, but the other side of the ladder must then carry its complement, making a base pair. Thus, DNA can carry information in the form of two complementary strings of 'letters'. During cell division (the process by which our bodies grow, repair themselves and reproduce), DNA strands replicate themselves using cellular machinery that 'unzips' the molecule down the middle and adds new complementary bases and backbone units to reproduce the missing half of the original molecule.

6 Protein factories

DNA's genetic code manifests itself through proteins – complex multipurpose molecules that make up all of our bodies' complex structures, themselves made up of countless smaller units called 'amino acids'. During protein synthesis, the DNA molecule temporarily unzips so that its code can be copied to an intermediate single-strand molecule called 'ribonucleic acid' (RNA). The RNA code is then 'read' by cellular machines called 'ribosomes', three letters at a time, with each three-letter sequence, or 'codon', corresponding to a particular amino acid to be added onto the growing protein molecule. Separate codes indicate the start and end of a protein strand.

7 The central dogma

In 1958, Francis Crick stated the so-called 'central dogma' of molecular biology, a simple but important statement about the way that information flows within biological systems. Essentially, it states that the genetic code from DNA is replicated onto a strand of RNA, and that information is then used to create the protein molecule. In special cases, proteins can be synthesized directly from DNA, and molecules of RNA can replicate themselves or even form a basis for creation of a new DNA molecule. Information is never transferred 'out' from proteins.

8 Sex cells

So how does an organism end up inheriting just one set of genes from each parent? The answer lies in the sex cells (sperm and eggs), which are generated through a special form of cell division. Known as 'meiosis', this process creates cells that carry a single set of genes, shuffled and remixed out of the two pairs present in the parent's body cells. The mix of genes in the sex cells varies randomly and is not influenced by whether a particular allele is a dominant trait. Hence, recessive genes from grandparents, not expressed in either parent, can pop up unexpectedly in a third generation.

9 Genetics and evolution

DNA provides the mechanism by which the entire process of evolution can work. By randomly mixing genes from two parents, it can give rise to individuals that are more or less suited to their environment, and more or less likely to pass those genes on to the next generation. In addition, the DNA copying process is not perfect – minor errors and changes can sometimes occur (and be overlooked by the cell's built-in error-correction processes). Often these changes are harmless and undetectable in the phenotype, but sometimes they give rise to new traits that can harm an individual's chances of survival, or boost their chances of reproduction.

10 Origins of DNA

Explaining the complexity of the DNA molecule is a key problem in evolutionary biology. The chances of even a small chunk of the DNA helix assembling by chance are astronomically small, so selection pressures must have somehow been able to create generations of molecules that got closer in form to DNA. The most popular solution, known as the 'RNA world' theory, relies on the single-stranded RNA molecule's more limited ability to carry genetic code. Studies suggest that chemical energetics would have allowed the building blocks of RNA to remain stable once formed, while only allowing them to bond with certain other chemicals in specific ways. RNA chains with competing chemistries may therefore have been the first forms of life, replicating by cannibalizing or 'cooperating' with rival strands until the first complete strands of DNA appeared.

TALK LIKE A GENIUS

‘Rosalind Franklin’s big contribution to the structure story came from photos of DNA she took using a technique called X-ray crystallography. Basically, you shine a beam of X-rays through a material and they get diffracted or spread out as they pass through the gaps between molecules. Watson and Crick, through no real fault of their own, ended up being privy to various parts of her work that were being shared without her knowledge, and that helped them figure out the double helix.

You can argue the ethics of how that information was shared around, but the real shame of it is that, because her work hadn’t been formally published, Franklin wasn’t even cited in the discovery paper. And because she died in 1958, she missed out on the possibility of a Nobel prize since they don’t award them posthumously. Despite this, she’s become such an iconic figure for women in science that, these days, you could argue that Franklin is even more famous than Crick and Watson.’

WERE YOU A GENIUS?

1 TRUE – geneticists love this common fruit fly because it has only four chromosomes and can breed in 10 days, making it ideal for seeing how genetic changes propagate.

2 FALSE – the separation and random mixing of genes actually happens during the creation of sperm and eggs.

3 FALSE – the DNA rungs are made from chemical bases attached to a phosphate backbone. Amino acids are used in the manufacture of proteins.

4 TRUE – for example, the four letters of DNA can make 64 three-letter codons, but these are used to indicate just 20 separate amino acids; each can be coded for in two or more ways, ensuring redundancy.

5 FALSE – small amounts of DNA can always be found outside the nucleus, in the mitochondria or cellular power plants.

DNA’s four-letter genetic codes can produce almost infinite variety once transformed into proteins in living organisms.

Human genome

'What more powerful form of study of mankind could there be than to read our own instruction book?'

FRANCIS S. COLLINS

The human genome is the sum total of all the genetic information that makes up a typical human being – a sequence of more than three billion nucleic acid 'letters' that occur along the DNA strands that make up the various human chromosomes. Putting this information together has been an enormous technological challenge, but understanding what it all means is an even greater one; we've barely begun to get to grips with our genetic make-up.

The genome is a book with some three billion letters – imagine what we could do if we knew what it all meant...

1 Humans have 98 per cent of the same genes as chimpanzees, but across our own species we are 99.5 per cent identical.

TRUE / FALSE

2 Celera Genomics, the privately funded company that sequenced the human genome, appealed to the public for DNA donations.

TRUE / FALSE

3 Genome research has identified about 100,000 meaningful genes in our chromosomes – more than anybody expected.

TRUE / FALSE

4 Junk DNA is redundant genetic code separating individual genes on a chromosome.

TRUE / FALSE

5 Sequencing the first human genome took 13 years, but a similar task can be accomplished today in just a few hours.

TRUE / FALSE

TEN THINGS A GENIUS KNOWS

1 Structure of the genome

A typical human cell carries the vast majority of its genetic information within a protected central region called the 'nucleus'. Here, long strands of DNA are curled up on each other to form the relatively compact coils of the chromosomes. In total, there are 46 chromosomes: 22 matched pairs known as 'autosomes', and 2 sex chromosomes (a pair of 'X' chromosomes in females, an 'X' and a 'Y' in males). The length of the chromosomes varies significantly, from almost 249 million nucleic acid pairs in the longest 'chromosome 1' to 57 million on the Y sex chromosome. In addition, small cellular 'power plants' called 'mitochondria' have a small (16,569 pairs) DNA packet of their own.

2 Sequencing DNA

The basic technique for sequencing DNA, pioneered by Frederick Sanger and team in 1977, is complex, but worth knowing (if only so you can understand what those charts on forensics TV shows actually mean). Individual DNA coils are first massively replicated using the 'polymerase chain reaction'. The coils are then 'unzipped' by application of heat, and snipped into many smaller fragments using chemicals called 'primers' that target particular base sequences. The fragments are then split between four vessels, with chemicals added that bond to the different letters on the unzipped strands, reconstructing the original DNA 'ladder'. But each vessel also gets one modified chemical – a dye keyed to a single letter, which terminates the reconstruction process at that point. End result: a mix of dyed DNA strands of different lengths, all known to end on a particular letter. Each vessel's contents are now smeared out across a plastic film in four columns (one for each letter), using a technique called 'electrophoresis' that ensures smaller, lighter strands migrate further; reading the DNA sequence is then simply a matter of seeing which column has accumulated DNA at a particular point.

3 Automated sequencing

The problem with the Sanger method, and most alternatives, is simply that there's an awful lot of DNA to sequence, and if you split it up into short strands of a few dozen to a few hundred letters, somehow you have to put it back together again. Fortunately, this is something that computers are good at: one approach, called 'shotgun sequencing', chops identical strands of DNA in many different ways and sequences them individually. Because the strands have different lengths, the codes overlap and a computer can sort through and work out how it all fits together. Modern sequencing techniques automate the whole task in parallel operations that speed things up even further.

4 The race for the genome

Famously, the goal of mapping the complete human genome turned into something of a race. The Human Genome Project (HGP), formally launched in 1990 with the backing of many governments around the world, found itself in competition with the private company Celera Genomics, which piggybacked on publicly released HGP data and used newer, faster techniques to catch up, despite only being launched in 1998. The HGP published a rough draft in 2000 (with Celera following in 2001), and effective completion in 2003.

5 Genes and junk

One surprising result of human genome studies has been the discovery that vast amounts of DNA are apparently redundant. When a cell needs to 'read' DNA in order to manufacture proteins, it only reads a relatively short portion of the strand, known as an 'exon' – long sequences of DNA in between the exons (known as 'introns') are simply ignored and don't seem to have much use – hence the term 'junk DNA'. Geneticists, however, prefer to call it 'non-coding DNA', since no one can prove for certain that it doesn't have some hidden significance. Results from the HGP suggest that up to 90 per cent of our DNA is non-coding, a much larger proportion than in simpler organisms; one suggestion is that introns accumulate and grow as DNA becomes more complex.

6 DNA profiling

Some 99.9 per cent of the genome is identical in all humans, but the remainder shows substantial variation across the population, and is unique in each individual (aside from identical twins). DNA profiling or fingerprinting targets these highly variable areas or 'loci' on a chromosome and looks for 'short tandem repeats' (STRs) – places where the same DNA letter is repeated several times. A few per cent of the population may typically share a certain repeat pattern, but if you check enough independent loci and still find a match between two DNA samples, it's highly unlikely to be coincidence. Patterns of STRs at specific loci form the basis for DNA databases, and can also be used to identify close relations, both for criminal forensics and daytime talk shows.

7 Mapping ancestry

Unlike chromosomal DNA (which mixes elements inherited from both parents), the short strands of mitochondrial DNA (mtDNA) are normally inherited only from the mother. As a result, they remain almost unaltered from generation to generation, though they do gradually accumulate changes over time through random mutations that are themselves passed on. This makes mtDNA a powerful tool for mapping relationships, both among humans and between many different living species: as a rule of thumb, the closer the match in mtDNA, the more recently two individuals shared a common ancestor. The same principle can be applied to the Y chromosome, which is only inherited along the paternal line.

8 The mother of us all

Mitochondrial research has led to some remarkable discoveries – for example, every human being on Earth today has an unbroken maternal line of descent to a single woman, 'mitochondrial Eve', who lived in Africa roughly 100,000 to 150,000 years ago (though she certainly wasn't the only female around at the time, and it's not the case that others of the time don't have living descendants, just that they don't have unbroken maternal lines). The spread of humans around the world can be traced through Eve's 'daughters', later women who are the most recent maternal common ancestors of large populations. A paternal line of descent can also be traced to 'Y-chromosomal Adam' about 200,000 to 300,000 years ago.

9 Genetic diseases

Studies of the genome have confirmed that a huge number of diseases have a genetic component. More than 4,000 diseases are known to arise through the expression of a single faulty gene, and such diseases obey the same patterns of inheritance discovered by Gregor Mendel in the 1860s. Diseases on autosomal chromosomes may be either dominant (the gene is always expressed if it is present, so there is normally a 50 per cent chance of the disease arising in offspring), or recessive (the gene is only expressed if two copies are present, so there is usually a 25 per cent chance if both parents are 'carriers'). Diseases caused by mutated genes on the X or Y sex chromosomes, meanwhile, have more complex patterns of inheritance. For example, the 'X-linked recessive' blood-clotting disorder haemophilia is inherited without symptoms by females, but can later be expressed in their male offspring – as some descendants of Queen Victoria unfortunately discovered.

10 Copyrighted genes

One of the trickiest issues to arise from genome sequencing is the question of who owns the data. While the publicly funded HGP made all of its data freely available, Celera Genomics founder Craig Venter courted controversy with his attempts to patent large numbers of naturally occurring genes. The issue was partially resolved in 2000 when US President Bill Clinton announced that the entire human genome should be freely available (sending biotech stocks plunging in the process) and the first draft genome was published. However, ethical issues linger on when it comes to patenting modified, non-human or artificially created genes.

TALK LIKE A GENIUS

‘The first organism to have its entire genome sequenced, in 1995, was the *Haemophilus influenzae*, a simple bacteria with 1.8 million base letters on a single chromosome.’

‘For all the talk about sequencing the full human genome, researchers have actually only covered about 92 per cent of it. Tightly packed clumps of DNA called ‘heterochromatin’ – found on the tips of chromosomes and where pairs join together in the middle – are so hard to unravel that they’re still working on those today.’

‘The Human Genome Project cost about $3 billion, while Celera sequenced the genome for about one-tenth of that. The HGP had to work out a lot of the technology from first principles, whereas Celera had the advantage of starting late and borrowing a lot of work that had already been done. Today, thanks to advances in technology, you can sequence a full human genome for less than $1000 – and even more cheaply if you just want the exons!’

WERE YOU A GENIUS?

1 FALSE – this is comparing two different things. We have 98 per cent of the same genes as chimps, but share 100 per cent of our genes with all other humans. *Within* those genes, all the diversity of the human race comes down to about 0.5 per cent variation in actual genetic code.

2 FALSE – Celera sequenced the DNA of their founder Craig Venter, while the HGP used DNA from a number of anonymous volunteers.

3 FALSE –it seems our genome has only 19 to 20,000 protein-coding genes – far fewer than the million-plus we once expected.

4 TRUE – but it’s rude to call it ‘junk’, since geneticists aren’t yet entirely certain it has no effects.

5 TRUE – and thanks to automation and more powerful computers it will soon be possible to sequence genomes in less than an hour.

Reading the genome is a huge first step on the road to understanding, and even manipulating, our cellular make-up.

Human origins

'It is ... probable that Africa was formerly inhabited by extinct apes closely allied to the gorilla and chimpanzee; and ... it is somewhat more probable that our early progenitors lived on the African continent than elsewhere.'

CHARLES DARWIN

Modern humans are unique animals in several ways – compared to other primates we have relatively little body hair, an upright bipedal stance and a massively enlarged brain. An estimated 20 million DNA nucleotides (roughly 0.6 per cent of our genome) have mutated since our lineage diverged from that of chimps, resulting in a string of increasingly familiar-looking hominid ancestors. Yet the story of human origins is far from straightforward, and recent discoveries have revealed close relatives surviving until surprisingly recent times.

Where did *Homo sapiens* come from, who were our ancestors and where did all of our cousins go?

1 A small amount of DNA in modern humans is Neanderthal, indicating that our ancestors interbred successfully with them. However, the Neanderthal DNA is only present in non-Africans.

TRUE / FALSE

2 The 'aquatic ape' theory argues that our human ancestors experienced a period during which they spent a lot of time in the water.

TRUE / FALSE

3 Scientists used the discovery of Peking Man fossils in 1921 to argue that China was the likely origin of the human species.

TRUE / FALSE

4 Our earliest suspected bipedal relative is dated to about 7 million years ago, but it could be more closely related to chimps than to humans.

TRUE / FALSE

5 The size of hominid brains has increased steadily throughout our evolutionary history. Modern *Homo sapiens* has a brain about four times larger than that of our shared ancestor with chimps.

TRUE / FALSE

TEN THINGS A GENIUS KNOWS

1 Apes are our close cousins

Humans and our nearest living relatives, the great apes, form a taxonomic 'family': the Hominidae. Fossils with hominid features begin to appear in the African fossil record from around 20 million years ago (mya), but in scientific terms, the hominidae became a distinct family a few million years later when the ancestors of gibbons (the lesser apes) branched off on their own evolutionary pathway. Genetic and fossil evidence suggest that, thereafter, the orang utans went their own way around 10 million years ago, with gorillas branching off around 6.3 mya. Humans separated from the common ancestor of chimpanzees and bonobos about 4 million years ago. (All of these dates are the estimated last times when our species shared genetic material through interbreeding; in practice, the major lineages were starting to diverge from around 13 million years ago, or more).

2 Our species is the only one of its kind

Because *Homo sapiens* is the only present-day representative of our genus *Homo*, it's easy to assume that extinct species form a straight line of descent back to the ancestor that branched off from chimps. But the true story is rather more complex – the human family tree has many tangled branches rather than a single trunk, with countless cousins, uncles and aunts at various removes. Working out exactly how the known extinct hominids fit into the picture is tough – especially when fossil remains are rare and fragmentary, and DNA can only be extracted and analysed from the most recent 'sub-fossil' bones of species such as Neanderthals.

3 Lucy and the southern apes

Outside of *Homo*, the other broad genus of hominids are the australopithecines or 'southern apes'. They show various human-like characteristics, such as upright walking, but are thought to have had chimp-sized brains (showing that bipedalism probably evolved before increased intelligence). The best known is *Australopithecus afarensis*, known as 'Lucy', the famous fossil discovered in Hadar, Ethiopia, in 1974. Lucy's species are often thought to have been responsible for the renowned 'Laetoli footprints', a series of bipedal tracks left by three individuals on a bed of recently deposited volcanic ash about 3.6 mya. The relationship between *Homo* and *Australopithecus* is still hotly debated; were the southern apes our ancestors or our cousins?

4 Olduvai Gorge

Fossil finds suggest that East Africa was the hotbed of early hominid evolution – in particular, the region around the Great Rift Valley. The most famous fossil site of all is Tanzania's Olduvai Gorge, where palaeontologists, including the famous Louis and Mary Leakey, unearthed fossils from a succession of ancient hominids, including *Homo habilis* or 'handy man' from 1.9 mya, its near-contemporary australopithecine, *Paranthropus boisei*, and *Homo erectus* from 1.2 mya. Stone tools found here and elsewhere suggest that primitive tool manufacture probably started among the australopithecines, and advanced massively to the classic flint hand axe by *H. erectus*. In recent decades, the focus of the search for earlier hominids has shifted further north to Ethiopia.

5 Bipedalism

The ability to walk on two feet for extended periods of time was a key advance in hominid evolution, and there are many theories to explain what drove it. Australopithecines such as Lucy seem to have lived on borders between forests and open savannah – environments in which evolutionary pressures may have supported a trading-in of ancestral climbing abilities for the ability to see further and walk for longer periods in a more energy-efficient upright style. Other theories are that upright walking reduced exposure to heat and allowed bipeds to stay active at times when other species were forced to siesta, and even that bipedalism (and hair loss) were adaptations to a life spent in and out of the water catching fish. Whatever the trigger, the consequences of bipedalism were dramatic, freeing the forelimbs to make better use of tools and driving the development of opposable thumbs.

Simplified family tree of the genus *Homo*

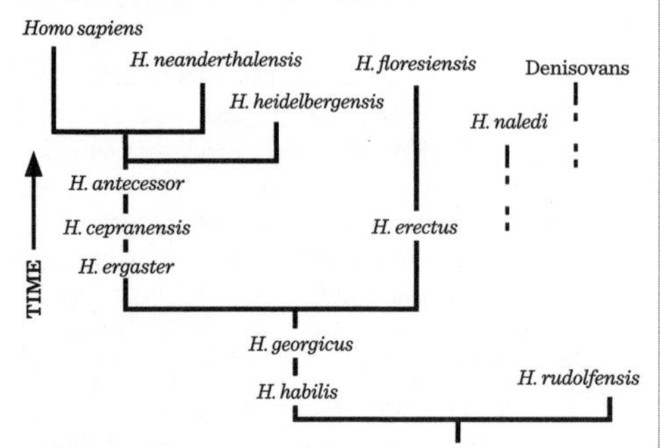

6 Out of Africa

Fossil and genetic evidence suggest that our ancestors attempted to spread around the world from Africa not once, but *three* times. *Homo erectus* was the first species to break out, some 2 mya, and succeeded in spreading across Europe and Asia. (The first *H. erectus* fossils were found in China; hence the species' early name of Peking Man). Different *erectus* branches ultimately gave rise to several new and distinct species, but modern humans are all descended from the branch that remained in their African homeland. Here, evolution gave rise to anatomically modern humans (*Homo sapiens*) at least 300,000 years ago.

7 Out of Africa II

Following in the footsteps of *erectus*, our own ancestors spread out of Africa from about 100,000 years ago. The latest evidence suggests that an initial *Homo sapiens* breakout through North Africa around 120,000 years ago ended in extinction for the migrants, but a second attempt through the Arabian Peninsula met with more success. By 50,000 years ago, modern humans were established across Europe, Asia and Australia. Humans did not finally reach the Americas until a 'land bridge' opened from Siberia at the height of the last Ice Age, about 15,000 years ago.

8 The Neanderthals

Perhaps the most famous extinct human species, *Homo neanderthalensis* once roamed widely across Eurasia. Unfairly characterized for a long time as misshapen, primitive brutes, the Neanderthals (named after the river valley in Germany in which early fossils were found) actually had larger brains than modern humans, and there's a growing mass of evidence for a sophisticated Neanderthal culture that includes burial rituals, art and jewellery. Although their fossils are exclusively Eurasian, genetic evidence suggests Neanderthals actually became a separate population while still in sub-Saharan Africa up to half a million years ago, before spreading into Europe and becoming extinct in their homelands about 160,000 years ago.

9 Newly discovered hominids

In recent years, archaeological finds have shown that some other hominid species survived until relatively recently. The Indonesian island of Flores was home to *Homo floresiensis*, a group of dwarf humans inevitably nicknamed 'hobbits'. They seem to have been descended from an *erectus* population that evolved in isolation for a million years or more, and they may have survived until as recently as 50,000 years ago. In 2010, analysis of DNA from a 40,000-year-old finger bone found in Siberia's remote Altai Mountains revealed another new species that we call the Denisovans. Perhaps most amazing of all, though, is *Homo naledi*, a hominid with a mix of ancient features, such as a small skull, and modern ones such as bipedalism. Found in South Africa and less than 250,000 years old, they were cousins of modern humans rather than direct ancestors.

10 Why *Homo sapiens* triumphed

With fossils suggesting the last Neanderthals clung on in the extreme southwest of Europe until just 40,000 years ago, it seems that three hominid species disappeared quite soon after *Homo sapiens* started its own journey out of Africa, so it's hard not to wonder if we were to blame. So far, however, there's no undisputed archaeology showing violence between rival hominid species. It's possible that our ancestors were simply better suited to a changing environment, perhaps out-competing their cousins for scarce food resources; certainly the arrival of *H. sapiens* in Europe coincided with a long period of significant climate change in the last Ice Age, and the final Neanderthal extinction coincides very closely with the last major cold snap.

TALK LIKE A GENIUS

‘ Almost every time someone discovers a new fossil hominid, they understandably try to show that it's one of our direct ancestors – who wants to dig up an extinct aunt or uncle? ’

‘ The view of Neanderthals as grunting troglodytes is wildly out of date, but it's been that way for 60 years now! It's mostly based on a male skeleton that was found at a French village called La Chapelle-aux-Saints in 1908. A palaeontologist called Marcellin Boule analysed it and commissioned illustrations that showed a stooping ape-man. What Boule missed was the fact that the skeleton was a 40-year-old (pretty ancient for a Neanderthal) who'd lost his teeth and had advanced arthritis. Researchers spotted the problem in 1957 and since then, scientists have realized that Neanderthals were actually very close to modern humans. But they still can't seem to shake off that old image – I blame Hollywood! ’

‘ There's an interesting theory that modern humans went through a ‘genetic bottleneck’ about 70,000 years ago when our ancestors were reduced to just a few thousand breeding pairs. The same thing seems to have happened to some other animals, and some researchers have linked it to the Toba supervolcano that erupted in Sumatra around that time. It's definitely still just a theory, though... ’

WERE YOU A GENIUS?

1 TRUE – Europeans, Asians and other non-Africans all carry about two per cent Neanderthal DNA.

2 TRUE – the aquatic ape theory suggests that our bipedalism, hairlessness and subcutaneous fat are all adaptations to a lifestyle hunting in water. However, the idea is very controversial.

3 FALSE –some scientists argued that Peking Man was the ancestor of Chinese people, but most remained focused on Africa.

4 TRUE – *Sahelanthropus* is an early hominid found in Chad. Only skull fragments are known but the arrangement of its spinal cord suggests an upright posture.

5 FALSE – while the brain-size comparison is correct, both Neanderthals and early European *Homo sapiens* had brains that were larger, on average, than our own.

Our family tree is a lot more varied than we once thought it was, but our origins lie in Africa.

Language and consciousness

'The deep structure that expresses the meaning is common to all languages, so it is claimed, being a simple reflection of the forms of thought.'

NOAM CHOMSKY

Some of the biggest questions in human evolution surround the things that seem to make us human – language, complex reasoning and self-awareness. None of these abilities leave direct archaeological traces, so it's hard to know when exactly they arose, and how they influenced the distribution and lifestyle of *Homo sapiens* as we spread around the planet.

How and when did humans make the leap from instinctive animals to conscious beings, and was this leap intimately connected to the development of language?

1 Most biologists agree that human language originates in a completely different part of the brain from typical primate calls.

TRUE / FALSE

2 A recent study suggests that our hominid ancestors were already going through evolutionary changes leading to a wider range of vocalizations as early as 4.5 million years ago.

TRUE / FALSE

3 One popular way of testing self-awareness in other animals is to see whether they notice a spot of dye on their reflections in a mirror. To date only primates, dolphins and killer whales have passed the test.

TRUE / FALSE

4 Some scientists have argued that the diffuse nature of consciousness is a result of quantum mechanical processes harnessed by special organs inside our brain.

TRUE / FALSE

5 Although they differ in their interpretation of the detail, most experts in linguistics accept Noam Chomsky's basic idea of a 'universal grammar'.

TRUE / FALSE

TEN THINGS A GENIUS KNOWS

1 The 'great leap forward'

So far as fossil evidence can show, humans have been 'anatomically modern' for about 200,000 years – yet for most of that time there is little evidence for the things that really distinguish us from other animals. One key area where evidence can survive, however, is the uniquely human trait of art, and here the evidence suggests that something significant changed in our way of looking at the world about 40,000 years ago. This is the period in which the first deliberately sculpted figures are known from at sites in central Europe, and cave art first appeared in Indonesia. With earlier evidence for art limited to abstract scratchings and the apparent use of objects such as seashells for adornment, many experts now believe there was a profound change in human capabilities around the peak of the last Ice Age.

2 Increasing brain size

One obvious trait in human evolution is an increase in brain size. While early bipeds such as the famous Lucy had similar-sized brains to chimpanzees (about 450 cm^3/27 cu in), modern human brains are about three times larger. One theory is that bipedalism turned our ancestors into more efficient hunters consuming a wider range of foods than the mostly vegetarian australopithecines, and thus provided the raw materials for brain growth. Another possibility is that the appearance of the truly opposable thumb, with its ability to grip small objects (first seen in *Homo ergaster* about 2 mya) drove the development of brain areas able to take advantage of this new ability.

3 Imagination and technology

A key human ability is that of imagining things beyond our immediate experience. Thus an ancient hominid could 'see' the possibility for modifying a flint pebble into a sharp-edged tool with a few well-placed chips along the edge and, indeed, visualize a use for such a tool in the first place. Scientists believe the act of imagination requires retrieving memories of past experience and remixing them in new ways. Studies of animal imagination rarely show such capabilities; although the use of tools has turned out to be far more common than was once thought, most animal tools are found objects. Only chimpanzees, elephants and crows have so far revealed the ability to fashion a tool such as a stick so that it is better suited for a specific task.

4 The language advantage

Studies of animals show they mostly learn by imitation – infants copy their parents in a whole range of activities and, when it comes to finding food, novices copy experienced individuals. Most animals have to make do without true language, however, which offers a key evolutionary advantage in the ability to describe a task in words, and offer advice on what the novice is doing right or wrong. By accelerating the learning process, even primitive language could offer a huge survival advantage and therefore a strong evolutionary selection pressure would have supported its rapid spread through the human population once it first appeared.

5 Fossilized evidence for speech

Humans have a unique ability to make a wide range of sounds suited to carrying an even broader range of meaning. From a physiological point of view, this ability is mostly linked to the shape of soft tissues in the larynx and palate. Because such tissues do not fossilize, it's impossible to directly track the appearance of speech in our ancestors. By way of a proxy, some palaeoanthropologists have fallen back on the positioning of the hyoid, a loose bone that migrates during male puberty to form the Adam's apple, and which is said by some to play a key role in our range of vocalizations. Others regard this as something of a red herring; an undescended hyoid certainly doesn't impede female language ability, and recent finds show that the bone was similarly placed in Neanderthals, who are thought to have had only limited language skills.

6 Are we hardwired for language?

Linguist and philosopher Noam Chomsky (b. 1928) famously argues that language is a far more innate ability than learning by imitation would suggest. His landmark book *Syntactic Structures* (1957) revolutionized the study of linguistics and displaced earlier 'behaviourist' theories that children simply learn by copying their parents, as was once thought. Indeed, studies show that, throughout their formative years, kids develop a clear understanding of complex grammar and syntax without ever being directly exposed to such ideas. Chomsky suggests that humans have uniquely evolved so that we're born with a language blueprint, a so-called 'universal grammar' (UG) with basic rules that we can leverage to rapidly learn whatever languages we're exposed to in our early years.

7 Language vs communication

Aside from Chomsky, Steven Pinker (b. 1954) is the other well-known name in the field of evolutionary linguistics. His 1994 book *The Language Instinct* supports Chomsky's general thesis, but differs in one important respect. Chomsky regards UG as a feature that arose uniquely in humans through a small genetic mutation that happened to add linguistic capabilities to existing 'neural machinery' (and changed little since it first appeared), Pinker argues that language is just one particularly useful aspect of a broader set of animal communications skills, and was therefore able to arise through selection pressures acting over much longer periods.

8 The language gene

Pursuing Chomsky's basic idea that language is dependent on the hardware of our brains, evolutionary biologists have scoured the human genome in search of evidence for hereditary language ability. The most important genetic candidate so far is a gene called FOXP2, whose malfunction is linked to certain speech disorders, particularly grammatical problems. The gene influences both brain and lung development, and specifically seems to affect the ability of neurons in the brain to make new connections (handy for language learning). While FOXP2 is found in all mammals, the human form suggestively has two coding differences compared to that found in chimpanzees. And while it's wrong to regard FOXP2 as a unique 'gene for language' (since many other genes are also involved), the recent discovery that Neanderthals shared our form of the gene is certainly intriguing.

9 How consciousness fits in

Did the evolution of language automatically trigger other changes in our mental outlook such as a 'theory of mind' – that is, the understanding that other humans have their beliefs and desires of their own that may not coincide with ours – and self-awareness? Undeniably, linguistic ability has important roles to play in abstraction from the purely physical realm, the sharing of points of view with others, and the construction of our mental models of the world, but is that all there is to it? In the past, scientists have tended to shy away from such thorny questions, which have attracted their fair share of mystics, but there are some fascinating theories with at least some grounding.

10 When consciousness arose

Two radically different alternatives offer good talking points for the would-be genius. Dismissed by most mainstream psychologists, the 'bicameral mind' theory, put forward by Julian Jaynes in 1976, posits that, until as recently as 3,000 years ago, the division between left and right brain (see page 66) was considerably deeper than it is today. We only became truly self-aware when this division broke down, and this radical shift can be traced through the changing perspective of ancient religious and philosophical texts. (Richard Dawkins hedges his bets, describing this idea as either complete rubbish or consummate genius). Conversely, the more recent 'attention schema theory' (AST; also unproven) argues that the core of consciousness, that idea of some disembodied mental entity inside us, arose slowly during vertebrate evolution as a result of the need to prioritize handling of sensory information in different situations. According to AST, a primitive organ called the 'tectum' – the back part of the midbrain – deals with stuff that requires our immediate attention, while the more advanced cerebral cortex roves freely around the 'background information', shifting focus in line with our conscious interests.

TALK LIKE A GENIUS

❛ Chomsky's ideas inspired a couple of attempts to study how chimpanzees learn – a female called Washoe was raised like a human child and learned about 350 words of sign language as well as apparently being able to combine them in quite complex ways. On the other hand, a male called Nim Chimpsky, taught in stricter laboratory conditions, learned about 125 and only used fairly limited combinations. If chimps really show grammatical ability, of course, that suggests that Pinker's right about language evolving over a longer period. ❜

❛ The problem with all of this is that mental faculties themselves don't fossilize; that's why burials are a hot topic for palaeoanthropologists. Death rites are an obvious sign of culture and a reasonably complex society, and the earliest generally agreed burials of modern humans are from about 100,000 years ago. Neanderthals made deliberate burials about 50,000 years ago, though they might have just been copycats – the real game-changer will be if they can prove that *Homo naledi* were burying their dead a quarter of a million years ago. ❜

❛ If the attention schema theory turns out to be right, then we're looking at pretty much every vertebrate animal being self-aware to some degree... so I think I'll maybe go for the vegetarian option? ❜

WERE YOU A GENIUS?

1 TRUE – primate calls seem to originate in the primitive limbic system, while human language seems to be handled by the highly developed cerebral cortex (see page 65).

2 TRUE – scientists found that the skull of *Ardipithecus ramidus* has features that would have helped it make more varied sounds.

3 FALSE – other animals that have passed the 'mirror test' include elephants, pigs and magpies.

4 TRUE – the idea of 'quantum consciousness' is explored further on page 74.

5 FALSE – there's actually fierce debate about whether UG works as a theory, with field researchers eager to point out apparent exceptions to the 'rules' in little-known languages.

Human beings are born with mental tools for complex language, but we don't know if they're connected to our self-awareness.

Nature vs nurture

'Often it is not so much the kind of person a human is as the kind of situation in which he/ she finds themselves that determines how they will act.'

STANLEY MILGRAM

Just how much human behaviour stems from our genetic inheritance, and how much is determined by our environment and upbringing? That question lies at the heart of a debate whose importance ranges far beyond the biological sciences, since for many people, it subtly affects our view of how the world works with impacts ranging across social policy, economics and philosophy. It almost goes without saying, then, that any would-be genius will do well to inform themselves on the evidence.

Are some people just born with certain tendencies? And if they are, should we do anything about it?

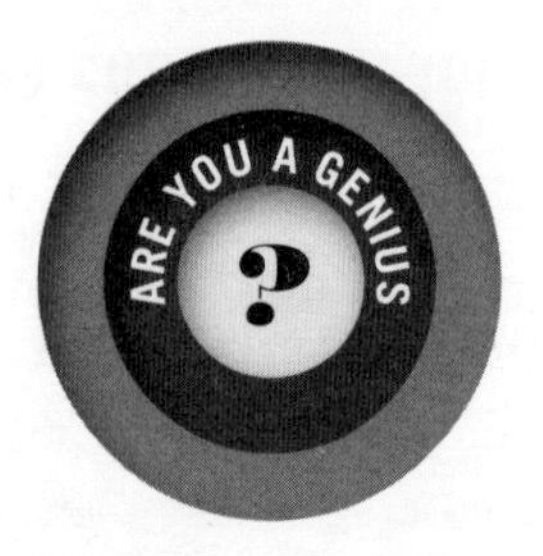

1 Although twin siblings frequently share similar political views, this is probably down to their shared upbringing, rather than genetic factors.

TRUE / FALSE

2 Biologists think that some inherited traits arise when environmental factors alter the genetic code of a person's DNA.

TRUE / FALSE

3 Despite a 'gay gene' having a clear disadvantage in natural selection, biologists think that genetic markers associated with male homosexuality could be carried by 50 per cent of the population.

TRUE / FALSE

4 Psychologist B.F. Skinner's experiments on chimps convinced him that we are primarily shaped by behavioural associations between stimuli and their results.

TRUE / FALSE

5 Although there is no specific intelligence gene, there's strong evidence that intelligence is at least partly an inherited trait.

TRUE / FALSE

TEN THINGS A GENIUS KNOWS

1 Survival of the fittest

Charles Darwin's theory of evolution by natural selection is undoubtedly one of the most powerful scientific tools for understanding the world ever discovered, but almost from the outset it's been misused and appropriated for political purposes. The term 'survival of the fittest' (an oversimplification of Darwin's thesis in the first place) has been used as a scientific fig leaf to justify everything from European imperialism through genocide to modern-day predatory business practice. Even today, mapping of the human genome threatens to establish new classes, discriminating between people based on their genetic inheritance and natural propensity to certain types of disease or behaviour. But how much are we really shaped by our genes, and how much by other factors?

2 Behaviourism and the blank slate

According to 17th-century philosopher John Locke, the mind of a newborn infant is empty of knowledge – a *tabula rasa*, or blank slate. The things we subsequently experience then fill up our memory and mould our attitudes as we grow. Locke's view was shared by many subsequent thinkers, including the behaviourist school of 20th-century psychologists. Influenced by the experiments of Ivan Pavlov (who trained dogs to salivate at the sound of a bell associated with food) and B.F. Skinner among others, they concluded that human learning was entirely a result of conditioning. By associating different behaviours or stimuli with various positive and negative outcomes, we develop a tendency to repeat certain behaviours and seek out certain stimuli while avoiding others.

3 Innate attitudes

The discovery of evidence for universal grammar (a set of rules for the construction of language that children are born with) in the 1950s began to undermine the purely behaviourist outlook. The unlocking of the human genome, meanwhile, has confirmed that many diseases (or the tendency towards them) are hereditary, raising obvious questions as to whether behavioural traits are also inherited through genes. Some experts argue that the pendulum has swung too far the other way, to a 'nativist' outlook in which our attributes are entirely innate, the result of variations in our deep genetic code. At the peak of this wave, around the turn of the century, it seemed as if barely a week could go by without scientists announcing the discovery of a new gene associated with disease or behaviour.

4 Twin studies

Studies of identical and non-identical twins offer a powerful tool for understanding how much we are shaped by heredity and how much by environment. You might think that the ideal twin study would be the rare case of siblings separated shortly after birth and raised in very different conditions, but a surprising amount can be learned by studying personality differences between twins raised in the same environment. What's more, those cases of separated twins that have been studied show that siblings frequently still share similar personalities and tastes. Some statistical analyses therefore suggest that about half of our personalities come down to genes (though we can't be sure of *which* half). However, there certainly isn't any sign of a generalized 'personality gene' yet.

5 Inherited intelligence

One obvious scientific question is whether children inherit intelligence from their parents? A popular 19th-century idea linked smarts crudely to brain size (a thesis that was not only grossly unfair to women, but when extended to various other aspects of skull morphology and behaviour, gave rise to the pernicious pseudoscience of eugenics). These days we know there's a lot more to intelligence than brain size, and there's no sign yet of a well-defined 'intelligence gene' (or indeed a mechanism for such a gene to act). The familiar IQ test attempts to measure innate intelligence (as distinct from the effects of a nurturing environment) and does seem to show that high IQs run in families, but it's far from perfect. And, intriguingly, people seem to be getting better at IQ tests over the generations – so are we all turning into geniuses?

6 Inherited adaptation

Surprising recent discoveries seem to show that our genes themselves can change, owing to environmental stress, and then be inherited by our offspring, a phenomenon called 'epigenetic inheritance'. The effect was discovered among the generations born after a World War II famine known as the 'Dutch Hunger Winter'. In the aftermath, studies showed that babies conceived in the famine but born afterwards 'caught up' to normal birth weight in the final trimester, but were more likely to suffer obesity and mental illness in later life. The curious thing is that these tendencies were inherited by their children in turn. One possible explanation is that exposure to the stress of famine modified the genome in both the foetus and its germ line (the cells that form the next generation of sex cells), but it's too soon to know if the phenomenon continues down the generations.

7 Epigenetics

The term 'epigenetics' is used to describe a whole range of changes to the phenotype (outward expression of the genes) that do not arise directly from changes to the genetic code. The way that genes express themselves can be modified by a wide variety of outside factors at different stages of life, blurring the lines between nature and nurture. Everything from changes in the mother's hormones during foetal development to infectious agents and perhaps even environmental pollution can change and disrupt the expression of genes. This may explain, or at least influence, conditions such as allergies, autism and obesity.

8 Sex and sexuality

One controversial area where the nature versus nurture debate plays out is the issue of sexuality. The campaign for LGBT+ equality has been aided by growing evidence that sexual preference is something we're born with rather than a result of the way we're raised. There's strong evidence that certain genetic markers are widely shared among, for example, gay men (though no single gay gene has been found). In general, this has been seen as a positive discovery, undermining outdated views of sexual preference as something that can be suppressed or changed by intervention. But while intolerant views linger on, can we be certain that efforts to discover such genetic markers might not be misused: for instance, to discriminate against LGBT+ people in certain societies?

9 Genes and criminality

An equally controversial question is whether genetics has a role in an individual's tendency to certain types of crime, and if so, how we should respond. For instance, statistical evidence links at least two genetic mutations to antisocial behaviour and low IQ in males (in particular, mutations in the 'MAOA gene' cause low levels of an enzyme that normally assists production of hormones that regulate body systems). Lawyers in the United States first tried to use a 'genetic defence' in a 1994 murder trial, but the first significant case of genetics being used in mitigation came in a 2009 Italian case. The link between genes and violent crime is certainly not simple, however: for instance, many individuals whose genetic mutation manifested in violent behaviour had also been victims of violence themselves as children, and no one has yet tried to suggest that either inherited abnormalities or abusive histories completely absolves an individual of responsibility.

10 Ethical issues

Both behaviourism and nativism raise some big ethical and philosophical problems: is it fair for society to reward individuals who happen to have been born into a nurturing environment that 'brings out the best' in them, or conversely to penalize those whose genetic make-up makes them less capable? And is there even such a thing as free will if we're slaves *either* to behavioural conditioning or to the particular proteins coded by our genome? A hard-line interpretation of either approach, if implemented across society, would likely lead to a dystopia of one sort or another, so most moderate people might hope that a balance can be struck. Fortunately, there's growing evidence that this is indeed the case: shaped by both genetics and our environment, our brains are fundamentally 'plastic', and capable of changing the way they function throughout our lives.

TALK LIKE A GENIUS

'In some ways epigenetics is rather reminiscent of Lamarckism, the old theory of evolution that was popular before Darwin. The idea was that parents could pass on characteristics developed during their life to their offspring, so giraffe necks got longer over time because each generation spent much of its time stretching to reach high branches. Obviously, the mechanisms in epigenetics are different, but it's funny how old ideas sometimes resurface in new forms.'

'And what about politics? Does genetics have anything to do with your general social outlook? There have been quite a few studies linking certain genes to whether people have conservative or liberal views, and there's particularly strong evidence around a dopamine receptor gene called DRD4. Dopamine's linked to our attitudes to risk, reward and punishment, so it's no surprise those are often a source of our fundamental political disagreements. But these studies usually suggest that a particular form of gene is only predictive if you combine it with environmental experiences like social interactions. So, as is often the case, it looks as if you can pin down about half of a particular tendency to genes; for the other half, you're back to nurture and upbringing.'

WERE YOU A GENIUS?

1 FALSE – studies of separately raised twins show a frequent match in political outlook, indicating that genetic factors are at work.

2 FALSE – so-called epigenetics doesn't change an individual's genetic code. It can, however, affect the particular genes that are *expressed*, sometimes over several generations.

3 TRUE – sisters of gay men (who share many of their genes) often have more children, so the genetic markers are spread widely even if they are not passed on by the men themselves.

4 FALSE – Skinner reached his conclusions largely on the basis of work with rats and pigeons.

5 TRUE – although it's hard to screen for effects of nurture and environment, studies of IQ tests suggest about half of our variation in intelligence is due to genetic factors.

Genetic inheritance plays a significant role in our behaviour and attitudes, but it rarely tells the whole story.

Genetic engineering

'Genetic engineering is to traditional crossbreeding what the nuclear bomb was to the sword.'

ANDREW KIMBRELL

Unlocking the secrets of the genome has opened the way for remarkable new techniques of genetic manipulation; no longer limited by traditional crossbreeding, we can now transplant useful characteristics wholesale between hugely different organisms. The potential applications are huge, but the creation of such genetically modified organisms (GMOs) also raises ethical and safety concerns. And the questions extend to human medicine; few object to the idea of locating genetic faults to track and treat hereditary diseases, but how far should we go when it comes to correcting the genes themselves?

Genetic engineering can transform our world and life itself – but how does it work, and what are the risks?

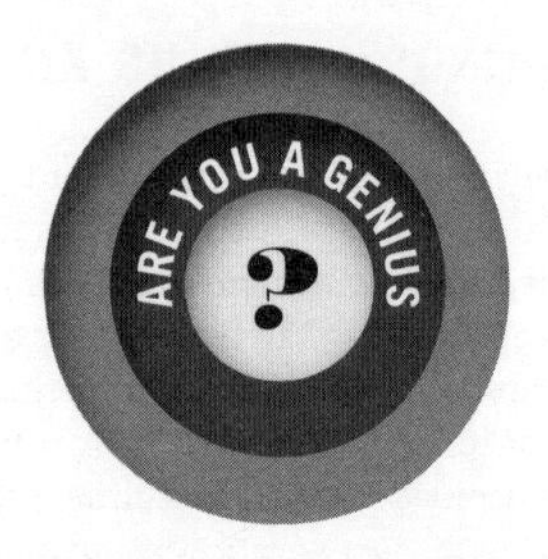

1 While genetically modified food is widespread in some parts of the world, genetically engineered drugs are banned from human use.

TRUE / FALSE

2 In 1997, scientists at the University of Massachusetts created a mouse with cartilage in the shape of a human ear on its back by transferring human genes to the genome of a naturally hairless mouse.

TRUE / FALSE

3 In 2014, scientists announced the creation of a plant that glows in the dark by transferring genes from bioluminescent marine bacteria into plant cells.

TRUE / FALSE

4 The first cloned mammal, Dolly the Sheep, lived to only 6.5 years, half the normal life expectancy of her breed, probably because the cloning process caused premature ageing.

TRUE / FALSE

5 Three-parent babies combine the genetic information from three people in their cell nuclei.

TRUE / FALSE

TEN THINGS A GENIUS KNOWS

1 Genetic engineering is ancient

Many people are instinctively cautious of the concepts around genetic engineering, seeing them as 'playing God'. But the reality is that our species has been intentionally tampering with the genetic make-up of other species since before the dawn of recorded history. Agriculture began (about 9500 BCE) with the realization by our hunter-gatherer ancestors that the seeds they ate could instead be replanted for harvesting in subsequent years, and selective breeding for plants with useful characteristics followed soon after. Around the same time, humans began the domestication and 'improvement' of various animals by selectively breeding those with desirable features. Like it or not, we live in a world that has been radically transformed by our modification of other organisms' genetic code.

2 First transgenic organisms

When most people think of genetic engineering, however, they imagine test-tube wielding scientists in white coats and sterile labs. In this sense, genetic engineering really began in 1972 when US biochemist Paul Berg found a way of splicing genes from two organisms together in a single strand of 'recombinant' DNA. Around the same time, Herbert Boyer and Stanley Cohen found a way of introducing foreign DNA into an organism that would then exhibit certain properties associated with the new DNA. The first 'transgenic' organism was a modest *E. coli* bacterium modified by a gene from another bacterium with natural resistance to certain antibiotics, but by 1974 the first transgenic animals were being created and the revolution had begun.

3 How to play God

Techniques for creating transgenic organisms vary from endearingly brutal to the downright crafty. At the direct end of the scale is a blunderbuss approach known as 'biolistics' or simply the 'gene gun' – a cellular-scale air pistol that fires tiny particles of gold coated with the desired DNA into a cluster of cells. Most of the cells are destroyed in the process, but a few survive and absorb the DNA into their own genomes. More ingenious are processes that harness the behaviour of viruses (which thrive by injecting their own DNA into living cells). The most common technique for creating genetically modified (GM) plants, meanwhile, utilizes *Agrobacterium*, a microbe that infects host plants with rings of DNA called 'plasmids'. Scientists first create a bacterium with a plasmid that carries the new DNA, and then use that to 'infect' embryonic plant cells

4 Benefits of genetic engineering

Within the lab, genetic modification has led to many medical breakthroughs – like it or not, the ability to create transgenic mice carrying genes that are linked to human diseases has vastly accelerated the pace of research in fields ranging from cancer to obesity and dementia. Experiments with transgenic fruit flies, meanwhile, have led to an improved understanding of how genes shape the development of our bodies and organs. The *E. coli* bacteria, meanwhile, has been engineered to produce a variety of products useful to the pharmaceutical industry, including human growth hormone, blood clotting factors and insulin.

5 Engineered animals

A variety of genetically engineered animals could also offer direct advances in medical treatment, though this opens up a variety of ethical issues (both in terms of taking the manipulation of animals so far, and in risks for the humans who would be likely beneficiaries). One of the most intriguing areas is xenotransplantation, the use of animal tissues and organs (usually from pigs) to replace failing human ones: genetic modification has experimentally been used to suppress production of 'antigens' that flag the transplanted organs as foreign and trigger rejection by the body's immune system (though so far the antigens always return after a while, so the treatment is not yet ready for medical use). Genetically modified animals may even have wider technological applications: in 2012, scientists at Utah State University unveiled transgenic 'spider goats' whose milk contains a protein that can be spun into a super strength spider silk with industrial applications.

6 GM crops

Perhaps the most controversial aspect of GM is its application to the food we eat. Transgenic crops engineered to be pest-resistant or drought-tolerant have become widespread in some parts of the world, but still raise alarm in others. While there is no evidence so far of GM crops being harmful to their consumers (and the technology theoretically offers huge opportunities to improve diets around the planet), a variety of concerns have been raised. These include the possibility of escape and unpredictable hybridization with wild crop strains, the risk of unforeseen knock-on effects to other parts of the environment, and regulatory worries about agribusiness companies forcing farmers to rely on their products.

7 Gene therapy

Scientists have long hailed the possibility of modifying human genes for medical treatment, but results have so far been limited. Humans suffer from more than 5,000 different genetic conditions, and gene therapy aims to replace the faulty genes with functional versions. The challenge lies in getting the modified genes to the tissues where they need to be; direct infusions of DNA into the bloodstream have limited success while modified viruses are effective in the laboratory but could present risks if they escaped into the environment. Another possibility tackles problems at source by modifying the genetic information in the germ line (sex cells) or in the fertilized embryo (creating a so-called 'three-parent' baby).

8 Clones and stem cells

The word 'cloning' naturally brings up science-fiction images of evil twins, but the reality is rather different. The production of cloned organisms involves bypassing fertilization by replacing the half-complement of chromosomes in an egg cell with a full set of chromosomes taken from another individual, and then kick-starting the development process. The first successful clone of an adult mammal was the famous Dolly the Sheep (born in 1996), but even today the process remains in its infancy, somewhat unpredictable and fraught with ethical issues. For instance, there are hopes of using cloning to harvest individually tailored 'stem cells' (cells capable of forming many different tissues during development) from early-stage cloned embryos, but the idea of creating embryos for therapeutic purposes is obviously controversial (happily, there are also genetic tricks capable of reactivating the potential of adult stem cells).

9 Overcoming ageing

Unlocking the secrets of the genome may also help to extend our lifespans and slow our ageing. Cells in the body replicate and replace themselves over a person's lifetime, a process that involves the splitting apart of paired duplicate chromosomes and construction of new counterparts before the nucleus and the cell as a whole can divide. However, each repetition of this process shortens protective 'caps' of non-coding DNA called 'telomeres' on the tips of each chromosome, and the process of senescence (when cells stop multiplying) seems to be triggered when the telomeres get down to a certain length. If it were possible to refresh these regions, could we extend our lives and even rejuvenate ourselves in the process?

10 Synthetic biology

Perhaps the ultimate form of genetic engineering is the creation of artificial life – cells with DNA code that is not borrowed from other organisms, but specifically created in a laboratory. In 2010, a team led by geneticist Craig Venter announced the creation of the first artificial life form, a *Mycoplasma* bacterium whose nuclear DNA had been replaced not with a genome directly taken from another organism, but with a modified copy constructed from scratch out of base chemicals. Scientists have been building up libraries of useful genetic code since the early 2000s, and can now create synthetic bacteria with metabolisms designed to produce fuel or digest pollution. The implications could be huge.

TALK LIKE A GENIUS

‘Many scientists would say that the only difference between genetic engineering and traditional crossbreeding is that you do one quite quickly in a laboratory and the other very slowly in a field. But maybe that's also at the heart of a lot of the concerns; when you crossbreed you're not talking about introducing a big genetic change very suddenly into a delicately balanced ecosystem.’

‘Synthetic biology has amazing potential – scientists have already come up with bacteria that can absorb carbon dioxide from the air and turn it into hydrocarbons, ones that start to glow in the presence of heavy metal pollution, and even ones that could potentially act as a universally compatible alternative to donated blood. The real problem is, what if someone uses the same technology to create a biological weapon?’

WERE YOU A GENIUS?

1 FALSE – GM drugs are widely used – for example insulin is now commonly manufactured using genetically modified bacteria.

2 FALSE – the 'earmouse' actually involved no genetic engineering at all; it was created by injecting cow cartilage cells into a biodegradable mould under the mouse's skin.

3 TRUE – pioneers hope that glowing plants could one day provide energy-efficient lighting.

4 FALSE – although Dolly died young, she succumbed to a lung condition common among sheep kept indoors.

5 FALSE – three-parent babies only have genetic material from two parents in their cell nuclei; only mitochondria from a donor are introduced to avoid disease risks.

Genetic engineering involves modifying the genome to produce cells and entire organisms with new characteristics.

The nature of reality

'Physical concepts are free creations of the human mind, and are not, however it may seem, uniquely determined by the external world.'

ALBERT EINSTEIN

Philosophers attempt to unravel the biggest mysteries in life, and there's none bigger than the question of reality itself. While we mostly get along fine taking the reality of the world around us for granted, that's not to say that our perception accurately tells us everything we need to know about it. The tangled question of how the world we perceive in our heads relates to objective reality (and whether such a thing even exists) has bothered philosophers both ancient and modern.

Figuring out where the realm of the mind stops and hard reality begins is a challenge that has tested even the greatest brains in history – can you do better?

1 The ancient Greek philosopher Leucippus was the first person to come up with a theory that the world is made from different kinds of indivisible atoms.

TRUE / FALSE

2 Plato's universal 'forms' represent an idealistic view of the world, in which objects only exist through our perception.

TRUE / FALSE

3 Thomas Hobbes used his materialist view of the world to argue against the existence of God.

TRUE / FALSE

4 Plato illustrated his idea of universal forms with a story known as the 'allegory of the cave'.

TRUE / FALSE

5 Common-sense realism has been undermined by the discovery of quantum physics.

TRUE / FALSE

TEN THINGS A GENIUS KNOWS

1 Monism

Many early philosophers held that our world is fundamentally made of a single substance that takes on the illusion of different appearances, or states – an outlook known today as 'monism'. For Thales of Miletus, who taught in the early sixth century CE, the base substance was water. His contemporary Anaximander believed that everything was made from an unidentified substance called *apeiron*, while a generation or so later Anaximenes saw air as the basis for everything. Many religious creation stories share a similar theme, with God repeatedly subdividing a primal substance to create the world. But these theories raised an obvious question – was the spirit made from the same substance, or was it somehow separate?

2 Heraclitus and the river of time

Heraclitus (*c.* 535–475 BCE, cuttingly nicknamed 'the Obscure' by his successors) is probably best known for the saying 'no man ever steps in the same river twice'. For him, the nature of reality was summed up in the inevitable flow of time, constant change and the unity of opposites. His idea was that all things are a balance between opposing properties that are continuously altering in relative proportion, so we only perceive the reality of a phenomenon by observing it at a certain time. In contrast, his contemporary Parmenides argued that not only was the world made from a single substance, but it was eternal and utterly unchanging – both variety and change are entirely illusory.

3 Plato's forms

Protagoras (*c.* 490–420 BCE) is the first person known to have made a living through philosophy as a professional 'sophist', but he is probably more famous for his insistence that 'man is the measure of all things,' a direct statement of the idea that individuals perceive different realities. Plato (*c.* 427–347 BCE), in contrast, insisted on the existence of an objective reality, but complicated things with his 'theory of forms'. In essence, this is the idea that the objects and phenomena we encounter in the everyday world are mere 'particulars' – imperfect copies of universal forms or ideals that have an independent existence outside of our minds in the 'world of forms'.

4 Aristotelean realism

Aristotle (384–322 BCE) took a different approach to the problem of universals and particulars. While Plato believed that individual objects and properties descended from or 'instantiated' universals that existed in their own world, Aristotle believed that the universals existed *within* the particulars, so that knowledge of universals could only be gained by study of particulars and the properties they have in common. He made meticulous studies in fields such as geology and nature, from which he derived theories about how the world works – the roots of 'natural philosophy' and modern science. The division between Plato and Aristotle's approaches created a deep division in philosophical schools that lasted until the Enlightenment of the 18th century.

5 Spirit and flesh

Medieval thinkers had very different ideas about the nature of reality from the classical philosophers. The Neoplatonist school influenced both Christianity and Islam, not least for the suggestion of its founder Plotinus (third century CE) that Plato's abstract forms originated as thoughts in the mind of a God-like 'One'. However Christianity's central mystery of a God incarnate on Earth also raised questions about the relationships of spirit and flesh, God and creation. Fifth-century Neoplatonists such as Proclus and the mysterious Pseudo-Dionysius developed ideas that echoed the Christian Trinity, but some religious thinkers took a different approach.

6 Dualism

Christian mystics, who saw a benevolent God as one side of an eternal power struggle with an equal, but opposite, force of evil, are generally know as dualists. By interpreting our world as the realm of the devil, they effectively let God off the hook for problems of evil and suffering, at the cost of reducing his omnipotence. Such heresies were ruthlessly put down by the mainstream church, although religious authorities were more accepting of the idea of dualism between the physical realm and that of the mind. Both of these approaches to duality can be found in other religions and philosophies; for instance, the ancient Persian religion of Zoroastrianism sees the world in terms of a struggle between good and evil principles. Chinese philosophies such as Confucianism and Daoism, meanwhile, see the world as a realm of balanced but opposing forces, yin and yang (reminiscent of Heraclitus), but rarely attribute them with fixed moral qualities.

7 The materialist Universe

English philosopher Thomas Hobbes made a decisive turn against the idea of a separate spiritual realm in his *Leviathan* (1651). Though mostly concerned with politics and government, Hobbes grounded his work in a fundamentally materialist view of reality; daringly, he assumed that there was no such thing as the independent soul, except where such a phenomenon arises from biological processes within the body: 'For what is the heart, but a spring; and the nerves, but so many strings...?' The scientific breakthroughs of the 17th century, from the circulation of the blood to Newton's laws of motion, created a growing sense that perhaps every aspect of the Universe might ultimately reveal itself to be some part of an intricate cosmic clockwork.

8 Perception and idealism

The 18th century saw new approaches to reality develop, rooted in the scepticism expressed by René Descartes as early as 1637 (see page 45). Descartes had argued that we can only really be sure of our own mental existence, and Irish philosopher George Berkeley's (1685–1753) 'subjective idealism' therefore argued that we conjure up the reality of the objects we perceive in our immediate experience; in effect, the Platonic forms lie dormant in our minds, waiting to be imprinted upon the world when we observe it. As a Christian bishop, Berkeley still allowed for the idea of a semi-objective external reality by privileging an omniscient God as the ultimate observer (capable of keeping reality going when we're not looking at it). His idea, an early form of 'phenomenalism', went on to influence thinkers such as Immanuel Kant and John Stuart Mill.

9 Common sense realism

Scotland was a major source of both intellectual and practical breakthroughs in the 18th and 19th centuries, but many philosophers of this Scottish Enlightenment found a contradiction between scientific and technological advances founded on observations of the 'real' world, and the phenomenalist approach that our perceptions create reality. Out of this emerged a school of 'common-sense realism', with Thomas Reid (1710–96) as its leading thinker. The Scottish realists essentially argued that things we perceive are objectively real, and we are made in such a way that our interactions with such objects create perceptions in our minds and help us form beliefs about them. The common-sense approach had a huge influence, especially among the Founding Fathers of the United States.

10 The simulation hypothesis

Technological breakthroughs of the 21st century (and influential sci-fi movie *The Matrix*) have raised a modern version of a question that was first asked by René Descartes in the 17th century (and is covered in detail in the next chapter). How can we be sure that the physical world is real at all? What if we're all victims of some 'deceiving demon', an advanced intelligence feeding our brains with virtual-reality stimuli while they ferment in a nutrient soup? We could even be lines of code running in some massive computer simulation with no physical presence whatsoever. We might sniff at such notions, but some philosophers have considered the possibility that we are actually part of a Universe in which a posthuman civilization has developed sufficient technology and interest to run such 'ancestor simulations'. In such a case, the number of so-called 'sims' experiencing apparent consciousness would be likely to vastly outweigh the number of actual conscious individuals in history – and we wouldn't know the difference.

TALK LIKE A GENIUS

“Common-sense realism is very much the kind of philosophy that the man in the street might come up with if you asked for his opinion on the nature of reality – but is it any the worse for that? Maybe all the more complex ideas are just a result of overthinking it...”

“A Greek philosopher called Leucippus probably came up with the idea of atoms, but he thought they were all made from a single substance rather than 118 different elements. Of course, if you drill down further you find that all atoms are made of protons, neutrons and electrons, but even the cutting edge of particle physics won't get you down to a single type of object yet. If you're looking for a monist interpretation of the Universe these days, then string theory might be your best bet.”

DUALISM

P M

P = Physical
M = Mental

MONISM

MATERIALISM P M

IDEALISM P M

SUBSTANCE MONISM P M

WERE YOU A GENIUS?

1 FALSE – Greek atomic theory involved just *one* type of atom, not the many different ones we know today.

2 FALSE – Plato's worldview is actually a type of realism, since he argued that both universals and the particulars that reflect their properties are objectively real.

3 FALSE – Hobbes' beliefs led to him being accused of atheism, but in fact he just had a different view of God from that of mainstream Christianity.

4 TRUE – Plato describes a cave in which chained prisoners can see only the shadows cast by objects passing in front of a fire that burns behind them. When one prisoner escapes, he is astounded to learn that this is not the true nature of the world.

5 TRUE – quantum physics (in which the properties of particles are only resolved when observed) does indeed undermine 'naïve' realism.

Is there such a thing as objective reality? Or do our interactions and observations of the world create reality around us?

I think, therefore I am

'And as I observed that this truth, *I think, therefore I am,* was so certain... I concluded that I might, without scruple, accept it as the first principle of the philosophy of which I was in search.'

RENÉ DESCARTES

Perhaps the most famous quote in all philosophy is also one of the most misunderstood. Descartes' renowned statement of certainty is, in fact, rooted in doubt about everything else. He derived it as the foundation of an ambitious project to build the methods of philosophy and knowledge about the Universe from first principles by thought alone – a 'rationalist' approach that was subsequently followed by many other philosophers.

If you're certain of your own existence, can smart thinking help you build a broader understanding of reality from there?

1 René Descartes believed that only the existence of our minds could truly be proved.

TRUE / FALSE

2 The Socratic method of establishing truth involves pitting a thesis against another argument called its antithesis.

TRUE / FALSE

3 Immanuel Kant defined the noumenon as the aspect of a thing that we can experience using our senses.

TRUE / FALSE

4 An *a posteriori* statement is one that describes observed facts about the world.

TRUE / FALSE

5 The statement 'the Eiffel Tower is 300 metres tall' would be defined by Leibniz as a 'truth of reasoning'.

TRUE / FALSE

TEN THINGS A GENIUS KNOWS

1 Origins of rationalism

Rationalism may be the oldest philosophy of all, with roots in the Pythagorean Greek school that flourished in the fifth and sixth centuries BCE. Pythagoreans believed that numbers had a mystical importance as the origin of all things, which led them to place great significance on the process of mathematical proof. A key question, of course, is how and why we have concepts of numbers and mathematics at all, to which rationalists answered that certain concepts and truths are innate; we don't need to investigate or discover them, we just *know* them intuitively. By then applying a process of logical deduction to these intuitions, we can further expand our knowledge. Subsequent Greek philosophers, including Socrates, Plato and Aristotle, all made use of these basic rationalist principles.

2 Socratic dialogues

The ideas of Socrates (*c.* 470–399 BCE) have come down to us only through the accounts of later writers (most significantly his pupil Plato). Many of Plato's works encapsulate the Socratic method – an approach to establishing knowledge based on logical sparring between people holding opposing positions. A typical Socratic dialogue might involve Arthur putting forward a statement (the 'thesis'), which Beatrice disputes by putting forward her own arguments ('premises') and hoping to persuade Arthur of their validity. Beatrice then shows Arthur that the premises he accepts contradict his original statement, which is thus either refuted or refined.

3 Avicenna's approach

Persian philosopher Ibn Sina (known in the West as Avicenna) was a key rationalist thinker of the early 11th century, around the height of the so-called Islamic Golden Age. He was convinced that reason and logic could be used to prove the truth of the Qu'ran, but is particularly well known for an insight that prefigures the ideas of Descartes. He imagined a 'floating man', suspended in the air and blindfolded so that he is robbed of his senses. Despite his sensory deprivation, the man is still aware that his mind exists, and has no physical substance of its own. For Avicenna, this was an insight into the immaterial realm of the immortal soul.

4 *Cogito, ergo sum*

Five centuries later, René Descartes (1596–1650) set out to develop a fundamental set of principles that could be used to investigate the Universe through rational thought. In doing so, he turned to 'methodological scepticism', an extreme position that discards any assumptions that can possibly be doubted in the hope of establishing true certainties. Descartes therefore dismissed all the evidence of his senses as potentially illusory or unreliable, but realized that, however extreme his doubts about the nature of existence, he could always be sure there was some entity doing that doubting. From this foundation, he was rapidly able to construct a belief in God, a certainty that God would not in fact, be deceiving him, and therefore a belief in the reality of the physical world. Descartes first published his insight in the French edition of his *Discourse on the Method* as '*je pense, donc je suis*', but later gave it a pithy Latin tag: '*cogito, ergo sum*'. Today, philosophers know it simply as 'the *cogito*'.

5 Deductive reasoning

One of Descartes' most important contributions is the method of 'deductive reasoning' – breaking down a problem under consideration into individual elements that have certain logical relationships and concluding whether the whole is valid. A classic deductive argument known as a 'syllogism' takes two conditional statements and combines them to produce a third (If I'm late, I'll miss the train; if I miss the train, I'll have to drive; therefore, if I'm late, I'll have to drive). Deductive reasoning is always logically watertight, provided the initial statements and terms involved are well-defined, but the method is limited when considering things you don't already know much about.

6 Spinoza's single substance

Descartes inspired many followers in the rationalist tradition, including 17th-century greats Baruch Spinoza and Gottfried Leibniz. Both had mathematical inclinations that drew them to adopt Descartes' method, but they differed in their approach to a key question of rationalism: the division between mind and body. Spinoza was particularly concerned that Cartesian dualism, the argument that body and mind are made of entirely different substances, rendered the physical Universe entirely mechanistic and denied a role for God to intervene. His solution, published posthumously in his *Ethics* (1677), was to effectively argue that God *is* the Universe, a single infinite substance out of which both physical and mental realms can be manifested. Spinoza's ideas are often regarded as progenitors to the school of 'neutral monism' developed in the late 19th century by physicist and philosopher Ernst Mach, and others.

7 Leibniz's rationalism

In contrast to Spinoza, Leibniz followed all aspects of Descartes' rationalism, including mind-body dualism (see page 73). One of his major contributions to rationalist thought is the idea that there are two kinds of truth: those of fact and those of reasoning. Truths of fact are necessarily correct by virtue of the concepts *within* them (for instance 'a triangle has three sides' is true because the *very definition* of a triangle, contained within the concept, is a three-sided figure). Truths of reasoning are *contingently* correct, since they rely on concepts from outside the statement itself (for example, the statement 'water boils at 100°C' is only true *because* we separately define the boiling point of water as 100 degrees on the Celsius scale).

8 Hume's fork

Scottish philosopher David Hume (1711–76) suggested that all our supposedly innate insights are, in fact, derived from empirical observations of the world. In the famous argument later called 'Hume's fork', he first divided all knowledge along similar lines to Leibniz (calling necessary truths 'relations of ideas' and contingent truths about the wider world 'matters of fact'). He showed that, while relations of ideas can be used to prove other relations of ideas, they tell us nothing about the outside world unless supported by matters of fact. What's more, matters of fact themselves are inherently prone to uncertainty, thanks to the limitations of our senses and observations. Hume's sceptical position, then, is that matters of fact are useful but provisional, while relations of ideas are provable but useless.

9 Kant's *Critique of Pure Reason*

Perhaps the most influential book in Enlightenment philosophy, Immanuel Kant's 1781 work is an attempt to reconcile the rationalist view with the rival empiricist approach in which knowledge is derived from observation. Kant, too, worried at the distinction between types of knowledge, defining Leibniz's truths of fact and reasoning as analytic and synthetic propositions respectively. He further distinguished innate *a priori* knowledge from experience-based *a posteriori* knowledge. Combining these concepts, he showed that analytic propositions are necessarily *a priori*, but while synthetic propositions are usually *a posteriori*, there's nothing to prevent the possibility of a synthetic *a priori* proposition. The remainder of Kant's notoriously difficult *Critique* attempts to construct such statements (he argued that arithmetic and geometry both contain them; 8 + 3 = 11 would be an example), and build them into a framework of 'metaphysics'.

10 Phenomenology

Kant's *Critique* culminates with his doctrine of 'transcendental idealism'; a general division of the world into 'noumenon' (the thing-in-itself) and 'phenomenon' (the thing as we experience it). It argues that, because the noumenal world cannot be accessed except via the phenomenal, the world we know (and the knowledge we derive about it) is inevitably shaped by our experience. Philosophers spent the better part of a century worrying about the interaction between Kant's two worlds before Edmund Husserl (1859–1938) realized it might be better to concentrate on the *Lebenswelt* – the world of phenomena we *can* experience. Husserl's 'phenomenology' concentrated on problems such as the individual experience and consciousness, and branched in many directions through the early 20th century. Key themes include the ways in which psychology, past experience and social pressure, as well as simple sensory perception, shape our perception and interpretation of the world.

TALK LIKE A GENIUS

‘Descartes himself expanded rapidly from certainty of his own existence to a wide range of certainties about the external world, but if you take the *cogito* to its logical extreme, you can end up with solipsism, where you continue to doubt everything except your own consciousness. That's not a very healthy outlook on life, really, since you also have to assume everyone else is nothing more than a figment of your imagination.’

‘The modern twist on Descartes and Avicenna's basic idea is the "brain in a vat", dreamt up by American philosopher Gilbert Harman. The idea is that, if you were an isolated brain being kept in a vat of nutrients and wired up to receive sensory signals simulated in a supercomputer, once again you couldn't really be sure of anything except for your own consciousness. If you think the basic idea is plausible then it can send you down a rabbit hole of scepticism, since if you can't be sure that you're not a brain in a vat, then you have to doubt anything and everything you think you know that's based on the assumption of you *not* being a brain in a vat. And that's pretty much everything.’

WERE YOU A GENIUS?

1 FALSE – Descartes only saw the certainty of our existence as a first step in proving the reality of the wider world.

2 FALSE – the counter-arguments put forward in Socratic dialogue are known as premises.

3 FALSE – for Kant, the noumenon is the unreachable 'thing-in-itself' – we can only experience it through the aspect known as the phenomenon.

4 TRUE – *a posteriori* translates roughly as 'after the fact', and applies only to knowledge gained from observation.

5 TRUE – our measurement of the Eiffel Tower's height is contingent on our separate definition of the metre.

Rationalism is a school of thought that derives truths by thinking alone. At its heart lies the observation that, at the very least, there's someone doing the thinking.

The search for understanding

'I believe it is worthwhile trying to discover more about the world, even if this only teaches us how little we know.'

KARL POPPER

Human society relies on learning about the world, principally through establishing rules about the way in which it operates – for instance, in the relationship between cause and effect or the principle of action and reaction. Today, we do this principally through the practice of science (a development of earlier natural philosophy). But what are the rules by which we can take evidence from our experience of the world and derive hypotheses about its operation? Over the centuries, many philosophers have attempted to tackle this question.

What's the best way to learn about the world, and how certain can we be of what we think we know?

1 Aristotle developed his method of drawing conclusions from observation while studying at Plato's academy.

TRUE / FALSE

2 David Hume argued that the only way we can understand the world is through close observation of its content and behaviour.

TRUE / FALSE

3 Modern science is based on the idea that proposed hypotheses can always be proved through logical deduction from careful observation and experiment.

TRUE / FALSE

4 Philosophers still can't make their minds up over exactly how knowledge should be defined, let alone how it can be collected.

TRUE / FALSE

5 In the 1620s, Francis Bacon developed methods for linking cause and effect that continue to play an influential role in the modern scientific method.

TRUE / FALSE

TEN THINGS A GENIUS KNOWS

1 Epistemology

Philosophers today call the study of knowledge 'epistemology'. The term was only coined in the mid-19th century, by Scots writer J.F. Ferrier, but attempts to derive theories of knowledge go back as far as philosophy itself. Epistemology has always held an understandable fascination for philosophers, since it forms the foundation of their attempts to understand everything else. At its heart lies a sceptical approach to our interpretations of the world: we may think we know things, but how can we be sure? Socrates, the pivotal figure at the heart of Greek philosophy, is said to have claimed that the only thing he could be sure of was his own uncertainty.

2 Aristotle's natural philosophy

Aristotle was a pupil at Plato's famous Academy for some two decades, but developed a very different approach to the derivation of knowledge. While Socrates and Plato were mostly interested in ethical problems and human behaviour, Aristotle was fascinated by natural phenomena that were not so amenable to the Socratic method of deductive logic, statement and refutation. Aristotle's method was, therefore, to collect as much data as possible, consider different ways of categorizing that data, and form hypotheses to explain the patterns that emerged. While the back-and-forth 'deduction' approach of Socrates and Plato inspired later rationalists, Aristotle's concept of 'induction' would ultimately give rise to the 'empirical' tradition of philosophy.

3 The birth of empiricism

Aristotle's works were highly regarded by early Islamic philosophers, but were only rediscovered by Western scholars during the 12th century. English friar Roger Bacon did much to reintroduce the Aristotelian approach to knowledge, and his Jacobean namesake Francis Bacon is widely credited with introducing the prototype of a 'scientific method' in the 1620s. In 1690, John Locke's influential *Essay Concerning Human Understanding* renewed an old argument of Aristotle's that the human mind at birth is a blank slate or *tabula rasa*, which is only filled up and shaped by sensory experience and our inductive reasoning. Empiricism is often seen as a British rival to rationalist 'continental' philosophy, but inevitably the story's more complicated than that; Leibniz, in particular, engaged with Locke and acknowledged the importance of experience, but at the same time argued that our understanding of the abstract world of mathematics implies a place for innate knowledge.

4 Hume's scepticism

Scotsman David Hume is often seen as an arch empiricist, but in fact he made arguments that undermined both rationalist and empiricist approaches to knowledge. Hume's *Enquiry Concerning Human Understanding* (1748) argues that all supposedly innate knowledge attributed to reason is in reality empirical. But it then raises awkward questions about whether what we learn from observation can really be called knowledge at all. Just because we always witness certain events and phenomena in association with each other, does this mean they necessarily *always* happen that way? Can we be sure of the links we perceive between cause and effect? And can we be sure the operation of the Universe won't change in the future? This is known as the 'problem of induction'.

5 Hegel's dialectic

Dialectic was a form of logical debate popular in medieval European universities – a formalized evolution of rationalist Socratic dialogue whose structure echoes down to modern public speaking debates. Its principles were revived by German philosopher Georg Hegel in the early 19th century, in an attempt to reconcile rationalist and empiricist traditions. 'Hegelian dialectic' is often described as the proposal of a thesis, its opposition by an antithesis, and the resolution of the two into a synthesis. However, Hegel himself used somewhat more informative terms: the initial thesis is the 'abstract' (derived perhaps from rationalist, deductive reasoning), the antithesis is the 'negative' (evidence to negate the abstract, provided by empirical, real-world experience), and the synthesis the 'concrete' (an accurate description of the real world). While Hegelian dialectic was a gallant effort, it has also been widely criticized; it remains familiar mostly because it has been used and abused to support various political viewpoints.

6 Logic

The late 19th century saw breakthroughs in the generalization of mathematical reasoning for wider use. Around 1850, English mathematician George Boole set out a system of symbols that allowed logical concepts to be treated mathematically, and in 1879 German philosopher Gottlob Frege published his *Begriffsschrift* (*Concept-Script*), showing how arithmetic emerges from basic 'first-order' logic. Frege's ideas caught on after Italian Giuseppe Peano and England's Bertrand Russell showed how they could revolutionize our understanding of particular fields of mathematics (see page 149). Their work gave rise to an influential school of 'analytic philosophy', and symbolic logic remains a powerful way of thinking about a range of philosophical, mathematical and linguistic problems.

7 The modern scientific method

Although Francis Bacon was the first person to propose schemes for securing knowledge from empirical observation in his 1620 *Novum Organum*, the modern scientific method owes more to the ideas of American philosopher Charles Sanders Peirce in the late 19th century. Peirce outlined a method of 'abductive' reasoning, the process we go through in order to reach a hypothesis based on observations. He emphasized the importance of 'curious circumstances', unusual observations that require some form of explanation to be reached intuitively. He also argued that in order to be useful, a hypothesis (once formed) must make some predictions about the world – in other words, it can be tested and proved or disproved.

8 Logical positivism

Influenced by Russell's work on logic, Wittgenstein's philosophy of language (see page 97) and the growing scope of science, an influential new school of philosophy arose between the two world wars. Largely based in Berlin and Vienna, the logical positivists believed in 'verificationism', the idea that statements are only 'cognitively meaningful' if they come from empirical observation and measurement (*a posteriori* statements, in Kant's terminology). In contrast, *a priori* propositions are 'cognitively meaningless' (they only carry meaning on their own terms because of how we define the language and concepts involved). This approach diminished whole fields of philosophy, such as ethics and metaphysics, reducing ethical choices to mere personal preference. However, the privileged position it granted to science meant that, unsurprisingly, it became very popular among scientists themselves.

9 The doctrine of falsifiability

Austrian-born Karl Popper (1902–94) took issue with logical positivism and developed his own highly influential counterpoint, placing great emphasis not on verification, but on 'falsifiability'. This is the idea that in order to be truly useful, any proposition must not only be cognitively meaningful, but also capable of being proven wrong (by a new and contradictory observation or measurement). Popper therefore acknowledged that classical inductive reasoning was not a path to absolute knowledge, but replaced Hume's problem of induction with a common-sense approach: an empirical theory can be assumed to be true *until such time* as some evidence appears to undermine it. This approach remains the *de facto* standard for most definitions of science.

10 What is knowledge?

For all our advances in learning, the definition of knowledge itself remains a central problem for epistemology. From the Enlightenment until the late 20th century, one popular argument stated that knowledge was 'justified true belief' (JTB); it consists of beliefs about the world that are both accurate, but also backed by evidence that helped us reach the conclusion. In 1963, however, American philosopher Edmund Gettier pointed out that the JTB definition was not really sound – it is possible to invent scenarios in which a belief is both justified *and* true, but is nevertheless just the result of a lucky guess rather than an accurate deduction, and should not really count as knowledge. In response to the 'Gettier problem', later epistemologists have tried either to reject his scenarios for lack of sufficient justification, or amend the conditions of JTB. Nevertheless, this definition of knowledge is the best we have for now.

TALK LIKE A GENIUS

“One of Popper’s favourite arguments against the positivist view is the case of black swans. The traditional view, he argued, was that you’d make a statement that all swans are white based on observation. But you can’t actually use deductive logic to prove that, just because all the swans you’ve seen are white, the statement “*all* swans are white” is true. Instead, you need to look at it the other way around; you make your hypothesis that all swans are white, and then when a black one turns up you *can* use deductive logic to prove that the statement “all swans are white” is false.”

“When talking about how science works, we can’t ignore Thomas Kuhn. His 1962 book *The Structure of Scientific Revolutions* upset a lot of people by overturning some cherished notions about steady scientific progress, and instead showing that scientists can be stubborn and resistant to change in the face of new evidence. Eventually, though, the weight of proof that the current theory is wrong becomes overwhelming and you get a sudden breakthrough to a new theory – a change of perspective that Kuhn called a ‘paradigm shift’. He may have exaggerated things a little, but the important central point is that scientists are more than just unthinking slaves to the scientific method.”

WERE YOU A GENIUS?

1 FALSE – Aristotle’s breakthroughs came from study of the wider world – specifically two years spent studying nature on the island of Lesbos.

2 FALSE – Hume actually doubted how much we can prove from either rational thought or observation.

3 FALSE – the modern scientific method is based on falsifiability – a theory is only ‘true’ until contradictory evidence appears.

4 TRUE – most definitions of knowledge, such as ‘justified true belief’, have some loopholes.

5 TRUE – Bacon’s method offers a technique for whittling down possible causes for a particular effect to the most probable, which can then be subjected to further investigation.

Theories based on observation and measurement are powerful tools for forming knowledge about the world – but we also need to remember their limits.

Good and evil

'Let me give you a definition of ethics: it is good to maintain and further life – it is bad to damage and destroy life.'

ALBERT SCHWEITZER

The concepts of morality and ethics seem to be an essential part of human existence. Individual definitions of right and wrong, good and evil, may vary widely, as do opinions about how they should be applied in practice, but even some of the most extreme attitudes and actions in history have been justified by people who see themselves, ultimately, as being in the right. It's little wonder, then, that philosophers have long sought to refine our definitions of right and wrong.

Can we justify our ideas about right and wrong, and why do we have such ideas in the first place?

1 Deontology is the belief that all ethics should be based on values that are ultimately taught by religion.

TRUE / FALSE

2 Epicurean philosophers believed that pleasure and pain were analogous to good and evil.

TRUE / FALSE

3 Jeremy Bentham's utilitarian calculus suggests that, sometimes, the interests of the few should be sacrificed for the benefit of the many.

TRUE / FALSE

4 Christianity inherited the idea of an 'original sin', derived from Adam and Eve in the Garden of Eden, from Judaism.

TRUE / FALSE

5 The 'golden rule' of morality is to do unto others as you would have them do unto you.

TRUE / FALSE

TEN THINGS A GENIUS KNOWS

1 The good life

The nature of virtue was a central concern of Greek philosophy, and thinkers from the time of Socrates onwards wrestled with the question of how it was to be attained. Most agreed it could be found through pursuit of '*eudaimonia*', the 'good life', but unsurprisingly they differed over exactly what this was. Socrates (as reported and later built upon by Plato) believed virtue was a largely internal pursuit – the cultivation of courage, self-control, justice and wisdom (but not necessarily putting them to work in the wider world). Both philosophers took issue with arguments that the achievement of wealth and earthly power was itself virtuous.

2 Aristotle's golden mean

Aristotle's *Nicomachean Ethics* elevated reason as the supreme and unique human characteristic, and therefore placed the exercise and cultivation of rationality at the heart of the good life, alongside the traditional virtues of Socrates. However, Aristotle also argued that *eudaimonia* was only to be achieved by exercising one's reason in the wider world – putting it to work and using it to moderate the more superficial virtues (preventing an excess of courage leading to recklessness, for example); what one really needs to achieve is a moderated middle way, the 'golden mean' between opposing vices. Going further, he argued that reason and virtue cannot achieve *eudaimonia* on their own – other external factors, such as health, relationships and even beauty, also play their parts.

3 Epicureanism and stoicism

Around 300 BCE, two different approaches to *eudaimonia* developed from the early ideas of Socrates, Plato and Aristotle. Epicurus and his followers took a sensory attitude that saw pleasure as the only intrinsic good, and pain as the only intrinsic evil. The good life is therefore one that pursues pleasure for both oneself and others, while avoiding pain and its infliction. In contrast, Zeno of Citium and his followers in the Stoic school returned to the Socratic view that the development of internal virtue was key to the good life. However, the Stoics differed in their conception of virtues, arguing that true virtue involves living in harmony with the natural world, and recognition of behaviours such as simplicity, self-discipline and honesty (similar to those in Eastern traditions such as Daoism and Buddhism). Stoicism became popular in a rising Roman Empire that sought to define itself in contrast to the fading power of classical Greece.

4 Where do virtues come from?

The Greeks may have been certain of the type of attributes they considered virtuous (and indeed, we've inherited most of those preconceptions ourselves), but there's an obvious question: *why* do we approve certain behaviours? Classically, the two key philosophical approaches to this question are 'deontology' (the assumption that ethics should be based on duties and obligations to society or a higher power), and 'teleology' (the argument that goals or outcomes are all that matters in judging ethical behaviour).

5 Christianity, Islam and sin

Monotheistic religions, such as Christianity and Islam, are unsurprisingly deontological in their approach to ethics, with sin defined as a deliberate breaching of God's various prohibitions, and virtue as obeying His laws (principally the Ten Commandments, inherited from Judaism in both religions). Christian theologians, however, still wrestled with the question as to *why* we sin, eventually coming up with the doctrine of 'original sin'. As described by Augustine of Hippo, this makes all humanity the inheritors of a damaging desire kindled when Adam and Eve ate the fruit in the Garden of Eden, and our sins can only be forgiven through the intercession of Christ. Islam, in contrast, denies the existence of original sin, but says that, while humans are born pure, they are also weak and forgetful; sins may be forgiven but it is every Muslim's duty to resist them.

6 Utilitarianism

Influenced in part by the materialist views of Thomas Hobbes (see page 42), the 18th-century Enlightenment saw a rebirth of interest in explaining ethics and prescribing morality outside a strictly religious framework. John Gay and David Hume both advanced teleological approaches, while the moral philosophy known as 'utilitarianism' is usually attributed to Jeremy Bentham. Writing around 1780, he argued that actions are morally justified if they promote the happiness of the greatest number of people while avoiding the infliction of pain on others. Bentham even developed a 'felicific calculus', a mathematical means for calculating the moral choice in specific situations (perhaps fortunately, it didn't catch on).

7 Kant and the golden rule

Reacting to utilitarianism, in 1785 Immanuel Kant published his own influential approach to morals and ethics. Arguing that man's possession of reason gave him a special place in the cosmos, he used that rationality as the basis for the 'categorical imperative' – the idea that you can only claim an action to be morally justified if you make decisions *as if* the maxims or principles behind it were to become the basis of a universal law. In some respects, however, this is just rebranding the principle of reciprocity, a 'golden rule' of morality that features in many world religions and philosophies. Monotheistic religions tend to give this rule a formula suggesting positive action: 'Do unto others as you would have them do unto you'. Other traditions emphasize restraint with a negative formulation: the Confucian 'Don't impose on others what you would not desire for yourself'.

8 Consequentialism

Utilitarianism remains a popular way of thinking about ethical decisions, as part of a broader 'consequentialist' approach. Consequentialism, as its name suggests, involves judging the morality of actions upon their consequences (an approach that appeals to many thinkers because it sidesteps the issue of defining the rules and duties of deontological morality). However, like all philosophers, consequentialists are prone to differences of opinion and a number of approaches have emerged. 'Ethical egoism', for example, puts personal consequences at the top of the list, while 'ethical altruism' does exactly the opposite, arguing that the individual should first consider the consequences of their actions for everybody else. 'Negative' consequentialism holds that avoidance of pain should take priority over promotion of happiness, while 'motive' consequentialism takes into account the intention of an action as well as its outcome. Finally, so-called 'rule consequentialism' is a hybrid system that argues that deontological rules and duties can be established based on their consequences.

9 'Boo-hurrah' and moral relativism

For modern philosophers, a big question is whether ethical statements are meaningful at all. This ultimately depends on the ideas of cognitive meaningfulness developed out of Wittgenstein's work on language in the early 20th century: if an ethical statement is meaningful, then it carries some real element of knowledge or truth about the world, while if it's meaningless, it can only express an emotional attitude. The concern goes back as far as David Hume, who in his 1739 *Treatise on Human Nature* noted the tendency of moral philosophers to jump suddenly from statements of what *is*, to what *ought to be* without properly bridging the gap. Two centuries later, British philosopher A.J. Ayer vividly framed the problem in his 'boo-hurrah' theory, arguing that ethical and moral propositions are little more than appeals for emotional censure or approval. In a similar vein, recent postmodern philosophy has led some towards 'moral relativism', the idea that any ethical system is merely a product of a certain set of rules we choose to apply, and that no particular set can justifiably be considered better than any other.

10 Ethics and evolution

Evolutionary anthropologists use Darwinian ideas to explain why we instinctively regard certain behaviours as ethical. In terms of simple evolutionary selection pressures, one might expect selfishness and self-preservation to be 'hardwired' into our characters, since they might improve our ability to pass genes and behaviour to the next generation. Why, then, do we find selfish and cowardly acts to be shameful (even if we can't always resist them)? Acts of altruism and self-sacrifice present even more of a puzzle. Many scientists link human ethics to the 'selfish gene' (the idea that behaviours that risk individuals can still propagate if such risks favour the kin group as a whole). Darwinian explanations for behaviours such as altruism towards other species, and why certain actions spark such strong feelings of approval or disapproval, remain elusive, or at least controversial.

TALK LIKE A GENIUS

‘Victorian liberal philosopher John Stuart Mill took Bentham's basic ideas of utilitarianism and felicific calculus further, arguing that the only justification for exercising power over individuals is to prevent harm to others. Unfortunately, he also argued that you had to distinguish between “higher” and “lower” pleasures to make sure the right sort of decisions got made. And of course you can bet that it would be high-minded people like Mill making the distinction!’

‘Diogenes of Sinope was one man who took Socrates' idea of rejecting worldly virtue to its extremes. He got rid of all his possessions and lived in a barrel wearing rags and picking arguments with passers-by. Athenians called him ‘the Dog’, but he certainly attracted quite a few followers who eventually took the insult doglike or *kynikos* as a badge of pride. They're better known as the Cynics.’

WERE YOU A GENIUS?

1 FALSE – deontology is only an acknowledgement that ethics derive from higher duties – but these may be to society rather than God.

2 TRUE – they therefore constructed a system of ethics based on the promotion of pleasure and avoidance of pain.

3 FALSE – Bentham argued that ethical actions promote the greatest good for the greatest number, but this did not extend to damaging the interests of minorities.

4 FALSE – most Jewish theologians believe that humans are born free of any inherited sin.

5 TRUE – and this rule appears in many forms in many different religions and philosophies.

We can justify our morality in various ways, but some would argue that we're really just pandering to our own emotions.

Free will and God

'Omnipotence and foreknowledge of God, I repeat, utterly destroy the doctrine of "free-will".'

MARTIN LUTHER

Do we really have absolute freedom in our actions, or are we merely hoodwinked by our perception of the world into *thinking* we have the liberty of independent decision making? Whether God allows free will and what that means for us have been crucial questions for theologians down the centuries, but the rise of science has brought a different perspective to the problem. Many modern philosophers argue that if you think things through, we really only have an *illusion* of free will. So where does that leave questions of ethics and morality?

Whether the Universe is ruled by an omnipotent God or follows the mindless clockwork of physics, is there any room for us to truly make our own decisions?

1 Ontological arguments for the existence of God are based on reason and logic rather than observations of the world.

TRUE / FALSE

2 The 'prime mover' argument is an ontological argument based on the idea that something external must have set the Universe in motion in the first place.

TRUE / FALSE

3 Compatibilism argues that we can't have free will in a deterministic Universe governed by cause and effect.

TRUE / FALSE

4 Spinoza argued that our ability to conceive of God must mean that He exists.

TRUE / FALSE

5 It's generally thought to be impossible to reconcile a belief in free will with a belief in an all-powerful God.

TRUE / FALSE

TEN THINGS A GENIUS KNOWS

1 Free will

The question of free will has been a central issue in Western philosophy since before the rise of Christianity. Ancient Greek religion included a strong belief that humans were mere playthings of the gods with irrevocable destinies, and while philosophers replaced the gods with laws of nature, these were no less implacable. Christianity complicated matters by raising questions about the afterlife. Other religions supported either a universal hereafter, specific fates for predefined elects, or (especially in the East) a cycle of rebirth,. Christians, however, found themselves torn between apparently incompatible concepts of predestination controlled by an omnipotent God, and personal salvation (the idea that acceptance of God and rejection of sin can guarantee you a place in the Kingdom of Heaven). How could God be omnipotent, yet still allow us to have free will?

2 Augustine and Boethius

In *On Rebuke and Grace*, Church father Augustine of Hippo resolved this problem through a different take on God's omnipotence. He suggested that people are born with both free will and a natural disposition to be good. Sin takes away free will, while profession of faith restores it. Predestination, in Augustine's view, arises because God already knows the future decisions we will make using our free will, and therefore who will ultimately be saved and who will not. Sixth-century philosopher Boethius, in *The Consolations of Philosophy*, clarified this argument by suggesting that God does not 'see the future', but actually exists outside time as we understand it. From this point on, the questions of free will and the existence of God become ever more inextricable.

3 The ontological argument

Eleventh-century monk Anselm of Canterbury put forward several arguments for God, the most famous of which is the 'ontological' argument or 'argument from existence'. This relies on the simple tenet that things that exist in reality are greater than those that exist only in the mind. For Anselm, therefore, God is the greatest being imaginable. Believers, and even atheists, can conceive of such a being existing, so it follows that it exists in the mental realm. But because a being that exists only in the mental realm would by definition *not* be the greatest conceivable being, God must exist beyond those confines in reality as well. Later figures, such as Descartes, Leibniz and mathematician Kurt Gödel, were all drawn to elaborate on this basic argument.

4 Islamic proofs of God

Persian scholar Avicenna (Ibn Sina) argued for the existence of God through his 'Proof of the Truthful' (*c.* 1035). He begins by considering the idea that things in this world are 'contingent' rather than 'necessary'; they only exist because some prior cause has made them exist. Since that prior cause was itself contingent, it too must have a prior cause, and so on. Avicenna concluded that because everything in the Universe is contingent, there must be a first cause *outside* the Universe, and that cause cannot be contingent, but must instead be necessary. Unsurprisingly, he then equated this necessary cause with Allah. A little over a century later, Andalusian polymath Averroes (Ibn Rushd) put forward a similar argument, reviving the Aristotelian idea that the motion of the Universe required a 'prime mover' to set it running. Averroes' case is an example of a 'teleological' argument, suggesting that features of the Universe show it must have been designed.

5 Negative theology

The idea that God is essentially unknowable beyond the fact of His existence lies at the heart of an apophatic or negative approach to theology that became popular in the Eastern Orthodox Church. Its best-known practitioner, however, is probably Córdoban rabbi Moses Maimonides, who described it in his *Guide for the Perplexed* (*c.* 1190). Maimonides argued that we should not make positive statements, such as 'God is powerful'; instead we should say 'God is not weak'. His chief arguments for the existence of God resemble those of Avicenna, but he added an important corollary regarding free will, namely that for God to be omniscient (all-knowing), humans cannot have freedom of action but must be compelled to follow the actions that God foresees.

6 The five proofs of Thomas Aquinas

In his *Summa Theologica*, 13th-century theologian Thomas Aquinas put forward five different arguments for the existence of God. These variously position God as the prime mover (in the physical movement of the Universe), the first cause (in a physical chain of cause and effect), the necessary end point to a chain of contingency (Avicenna's argument), and the architect of cosmic design. Aquinas's original contribution, however, is his fourth proof, the 'argument from degree'. This form of ontological argument states that, since all things have degrees of properties from lesser to greater, there must be a *greatest* possible quantity to such properties that is nowhere surpassed, and therefore God is the entity that has all properties in the greatest degree.

7 Spinoza's 'God *is* nature'

Baruch Spinoza's unique views led to his exile from the Dutch Jewish community in which he was born, and later saw his books forbidden by the Catholic Church. In his posthumously published *Ethics* of 1677, he develops a unique model of reality based on cause, effect and laws of nature, foreshadowing our modern scientific view of the Universe in many ways. Spinoza's ontological argument for God is broadly similar to Anselm's, with the twist that the human mind is incapable of forming ideas without external cause, so the fact that we can conceive of God must mean that He exists. Spinoza's conception of the deity, however, is very different from the one most people derive from scripture; he sees God as the sum of all natural laws, and therefore, ultimately, as the infinite substance out of which the Universe itself is made. Fiercely attacked at the time, Spinoza's views nevertheless set the stage for the 18th-century Enlightenment.

8 Determinism and compatibilism

The debate over free will was central to the religious Reformation of the 16th and 17th centuries. While the Catholic Church stuck broadly to the arguments of Augustine and Boethius, Protestants generally argued that free will was limited to certain spheres, with our spiritual fate entirely predetermined. Spinoza was perhaps the first modern philosopher to take an entirely deterministic view of the free will question; his ideas of Universal cause and effect meant that free will could not exist, because all our actions have prior (and ultimately external) causes leading back to God. Materialist Thomas Hobbes put forward a possible middle way in *Leviathan* (1651), arguing that we have 'freedom of action' in accordance with our will, but nevertheless our will is not free since it is shaped by external motivating factors. This is an early example of a 'compatibilist' approach, attempting to reconcile free will with the nature of the Universe (often by tightening the definitions of what 'free will' actually entails).

9 Atheism

From the Enlightenment onwards, the loosening of religious power allowed philosophers to speculate more freely on the nature of God and indeed whether God existed at all. Immanuel Kant's transcendental idealism (dividing the world between objective noumenon and perceived phenomenon) inspired a German idealist movement that dominated 19th-century philosophy. Its key work, Arthur Schopenhauer's *The World as Will and Representation* (1818) argued that noumenon and phenomenon are essentially the same thing – acts of will – viewed from within and without. He went on to construct a model of a Universe on compatibilist lines, openly at odds with traditional views of God and more in line with Eastern religions, such as Hinduism and Buddhism. This paved the way for later and more aggressively atheistic philosophers, such as Nietzsche.

10 Free will in the quantum Universe

Since the 19th century, the focus of the free will debate has switched from questions of God's omniscience to concerns about physical determinism. The idea that a clockwork Universe governed by immutable laws of physics suggests that our actions are never truly free, but are driven by a predetermined history that can be traced back to the Big Bang itself. Bothered by the implications of this, some philosophers and many scientists have found solace in the development of quantum physics (see page 174). At first glance, quantum ideas appear to at least offer a place for uncertainty, although randomness does not itself equate to us having more control over our decisions. However, several new compatibilist approaches have attempted to expand this into a possibility of true free will.

TALK LIKE A GENIUS

' From our point of view, though, does it really *matter* if we have free will, provided we think we do? Even if the Universe is entirely mechanistic and effect stems from cause all the way back to the Big Bang, our own decision-making processes are part of that. And provided *we* don't know what decisions we're going to make or feel forced into making, then perhaps it doesn't really matter if, in some sense, they're foreseen by the Universe? '

' Some psychologists are concerned about what might happen in a society that stopped believing in free will, and if you look at all those experiments where people are persuaded to act like bastards because they're "just obeying orders", I guess you might be worried. (Take the Milgram Experiment for example, where participants were ordered by an authority figure to inflict electric shocks on someone in a different room and mostly just went along with it.) But that's really talking about something different – free will or the lack of it shouldn't really impact on us taking moral responsibility for our actions. '

WERE YOU A GENIUS?

1 TRUE – Bertrand Russell argued that this is why many people find them unsatisfying, but also have difficulty actually disproving them.

2 TRUE – this kind of argument for the existence of God was popular during the Islamic golden age and in medieval Europe.

3 TRUE – however, although our will is always determined by the chain of cause and effect, compatibilism does allow for a lesser 'freedom of action'.

4 TRUE – Spinoza said that mental ideas required external causes.

5 FALSE – it's actually the all-knowingness of God that is the problem (though Augustine and Boethius both point out ways around this problem).

Free will is an idea we cherish, even though there are good reasons to think that it's probably just an illusion.

Existentialism

'I tore myself away from the safe comfort of certainties through my love for truth – and truth rewarded me.'

SIMONE DE BEAUVOIR

Existentialism has a reputation as the most dour and angst-ridden of all philosophical movements, yet in reality it is also the most profoundly human, originating as it does in a feeling that previous schools of philosophy had little to say about the human condition. At its heart, it is a belief that we are fundamentally free – life has no deeper meaning, and therefore we alone are responsible for our personal choices and destinies. This means not only embracing the feelings and actions of our everyday human lives, but sometimes also confronting our innermost anxieties.

Confronting the meaninglessness of existence can be a liberating experience – but are you ready for the challenge?

1 Friedrich Nietzsche presented compelling arguments that the Universe was not created by a God.

TRUE / FALSE

2 Martin Heidegger argued that pure rational thought is the only way in which we can come to terms with the lack of meaning to our existence.

TRUE / FALSE

3 Albert Camus said that the Greek hero Sisyphus would ultimately achieve happiness when he successfully pushed his rock to the top of the hill.

TRUE / FALSE

4 Authenticity is the existentialist idea that we must always present our true faces to the world.

TRUE / FALSE

5 Jean-Paul Sartre argued that the existentialist struggle involved coming to terms with the fact that we are doomed to a short time on Earth.

TRUE / FALSE

TEN THINGS A GENIUS KNOWS

1 The continental/analytic divide

If the long-standing divide between rationalism and empiricism had been bridged to some extent by Kant's *Critique of Pure Reason* (1781), a new split soon arose in its place. By the mid-19th century, many philosophers (particularly in English-speaking countries) were pursuing an analytic approach; dissatisfied with Kant's views about knowledge, they instead embarked on a project of building knowledge logically from the smallest possible propositions. Meanwhile, a rival school sprang from the ideas of Georg Hegel, who fundamentally rejected Kant's division of phenomenon and noumenon (the world-as-experienced and the unknowable reality) in favour of a single united 'Idea'. Hegel's thinking gave rise to German Idealism, which attempted to build philosophical ideas around a grand, overarching vision of existence. Idealism and its intellectual heirs eventually became known as the continental tradition.

2 Kierkegaard and Nietzsche

Two figures in 19th-century Europe did much to pave the way for existentialism. Danish theologian Søren Kierkegaard (1813-55) took a particularly keen interest in questions of how individuals should live their lives. This led him to write widely on topics such as faith, angst, ethics and passion. He argued above all that belief in God required a 'leap of faith', and that truth is to be found in subjectivity (it is our response to facts that matters, not the facts themselves). Friedrich Nietzsche (1844–1900) went much further. Influenced by Schopenhauer's ideas about the force of acts of will, he argued that 'God is dead' – a statement that rejected Christianity as a deontological, 'slavish' moral system that could no longer be supported. Nietzsche believed that in its place, humans should pursue self-actualization within a 'masterly' framework that judged morality in terms of consequences rather than duties.

3 Dostoyevsky and Kafka

Early existentialism is characterized by its appearance in the work of several major European authors. Fyodor Dostoyevsky's *Notes from Underground* (1864) is a startling insight into the mind of an alienated individual that focuses heavily on the anti-hero's interior monologue as he contemplates his place in the world and the nature of his existence, before a series of interactions with the outside world open up his consciousness. Dostoyevsky's deep psychological inquiry and broad scope foreshadow later works that had a far wider impact. In the early 20th century, meanwhile, Franz Kafka revisited many of Dostoyevsky's views on the apparent meaningless of existence, while adding his own particular blend of bleak absurdism.

4 Heidegger's *Being and Time*

If earlier continental philosophers had addressed specific questions about the nature of existence, German Martin Heidegger was the first to mould the question of being into a complete philosophical system for shaping our interactions with the world. *Being and Time* (1927) focuses principally on the question of what it means for us to exist in this world (something he called *Dasein*, or 'being there'). Where rationalists had seen thought alone as the defining action of human existence, Heidegger argued that caring, our unavoidable engagement with the concerns of the world, was much more important. Alongside this, he believed that we are fundamentally shaped by our relationship to time – the inexorable slipping away of our lifespan and narrowing down of possibilities. He argued that the only response to this is to live in a way that is authentic, or true to the inner self.

5 Jean-Paul Sartre

Probably the archetypal existential philosopher, French thinker Sartre explored his ideas through both politics and literature and became a dominant cultural figure in the mid-20th century. An avowed atheist, his 1943 book *Being and Nothingness* offers his philosophical manifesto. He takes the position that humans must define their own purpose in life, since 'existence precedes essence' (in other words, while our own creations may have purpose since we produce them with a use in mind, the lack of a creator God removes the possibility of there being a similar innate purpose to our own existence). Life experience is therefore the exercise of a deep-seated freedom manifested through action. Sartre also argues that realization of the fact that we are free (often resulting from a growing awareness of our finite lifespan) can be a deeply unsettling experience (hence the title of his earlier 1938 novel *La Nausée*).

6 Authenticity

The concept of authenticity originates in the writings of Kierkegaard, who was concerned with how individuals can develop an authentic religious faith in the face of both Church dogma and a view of the world inevitably skewed by media interpretations. Shorn of its religious overtones, authenticity re-emerged as a major issue for 20th-century existentialists, most notably Sartre, who saw it as an acceptance and embrace of one's inner needs and rejection of the strictures of bourgeois society. In their quest for authenticity, existentialists therefore frequently adopted bohemian lifestyles (famously exemplified in the open relationship between Sartre and Simone de Beauvoir) and anti-establishment positions.

7 Absurdism

One key theme often linked to existentialism is the absurd; the conflict that arises between the human tendency to seek meaning and purpose in the world and the existentialist belief that there is no such purpose to be found. Kierkegaard was among the first to wrestle with this problem, even going so far as to consider whether suicide was a viable means of escape. However, he ultimately concluded that a better solution was to make a 'leap of faith' embracing the religious idea that there is some higher entity or purpose despite the lack of rational evidence. A century later, philosophers such as Albert Camus viewed such a leap of faith as an intellectual surrender akin to suicide itself, and argued instead that the correct response is to embrace the absurd.

8 Camus and *Sisyphus*

Camus rejected the existentialist label, but nevertheless wrestled with similar issues around how we should live if our existence has no innate purpose. His 1942 essay *The Myth of Sisyphus* argues for revolt in the face of a meaningless Universe – embracing our existence in the moment, and our fleeting moments of happiness and fulfilment. It explores the absurdism to be found in various human pursuits, concluding with the myth of the Greek king Sisyphus, condemned by the Gods to forever roll a boulder up a hill, only for it to roll back down and strike him. The daily struggle of existence, for Camus, is similarly futile and laborious, yet this is not to say that we cannot find contentment in our acceptance of it.

9 Simone de Beauvoir

Best known today for *The Second Sex*, the 1949 book that remains a foundational text of the feminist movement, de Beauvoir was equally influential as an existentialist philosopher and novelist. Her 1947 book *The Ethics of Ambiguity* was both a response to her partner Sartre's *Being and Nothingness*, and an attempt to build a comprehensive ethical system on existential foundations. A lifelong atheist, de Beauvoir argues that human consciousness in a godless Universe gives us absolute freedom, but that freedom can only be expressed through 'projects', physical actions that put our free will to work. She identifies various human responses to our freedom, ranging from attempts at denial through surrender to other powers both real and imagined, to nihilistic despair and rash adventurism. Finally, de Beauvoir outlines a concept of moral action driven by the realization of our freedom through projects in the real world, balanced against a refusal to oppress others.

10 Existentialism in the arts

Aspects of existentialism manifested widely in society from the 1950s through to the present day (particularly during the Cold War era when the prospect of impending nuclear holocaust seemed an ever-present reminder of the absurd). Its basic ideas made their presence felt in the French 'New Wave' cinema and the American 'art house' movement, and the experiential 'gonzo' writing of authors such as Jack Kerouac and Hunter S. Thompson. In theatre, it inspired both the bleak outlook of Beckett and Pinter, and the absurdism of Eugène Ionesco and Tom Stoppard. Even debates over the 'authenticity' or otherwise of various popular musicians and styles from jazz to rock can trace their roots back to the existential debate about authenticity and conformity.

TALK LIKE A GENIUS

‘Nietzsche gets a bad press, mostly from people saying that his philosophy paved the way for Nazism – but that's actually pretty unfair. He was friends with Wagner from his school days, but broke off relations and published damning criticism after Wagner voiced support for anti-Semitism and the unification of all German-speaking countries in a new Empire. He ditched an editor for similar reasons. The problem was that after his mental collapse in 1889, his sister Elisabeth took control of his literary estate. She on the other hand *did* have a lot in common with Wagner's views, and started rewriting Nietzsche's unpublished *Will to Power* to support them. The truth only came out when scholars obtained access to Nietzsche's notebooks in the 1960s, but by that point the damage had been done.’

‘If you're looking for existential theatre that still has some decent jokes, then *Rosencrantz and Guildenstern are Dead* is your best bet. Tom Stoppard's 1966 play follows the misadventures of two bit-part characters from *Hamlet*, as they debate the nature of existence in between trying to figure out what the heck is going on from the bits they see of Shakespeare's plot.’

WERE YOU A GENIUS?

1 FALSE – Nietzsche was not concerned about the creation of the Universe, so much as the effects of religious morality on humankind.

2 FALSE – Heidegger actually argued that we can only overcome the lack of meaning by engaging and caring about the world.

3 FALSE – Camus says contentment only comes through acceptance of the pointless struggle of existence.

4 TRUE – Sartre contrasted authenticity with a dishonest ‘bad faith’ about our true selves.

5 FALSE – for Sartre the terrifying thing is not our short life spans, but the fact that we are free agents.

If the lack of a creator means that life has no purpose, then existentialism argues that we must embrace our terrifying freedom.

The human brain

'The human brain has 100 billion neurons... Sitting on your shoulders is the most complicated object in the known Universe.'

MICHIO KAKU

A mere 1.5 kg (3.3 lb) or so of fatty tissue rattling around in our skulls, the human brain is nevertheless capable of astounding feats, and a source of endless fascination to scientists and philosophers alike. Billions of individual cells form a network of at least 100 million *million* connections, making possible the entire panoply of human artistic and scientific achievement. Neuroscience, the field that investigates the brain's physical and mental properties, has revealed many surprising secrets, but our grey matter still conceals plenty of mysteries.

Being a genius is all about brain power, but can we untangle the mental mysteries that make us all, in our own ways, extraordinary?

1 Despite its amazing capabilities, the human brain consumes fewer energy resources per kilogram than the rest of your body.

TRUE / FALSE

2 The average brain is thought to have at least 100 trillion separate neural connections, shared roughly evenly between about 83 billion individual neurons.

TRUE / FALSE

3 Brain scanners trace activity through electrical emissions that are created as tiny sparks jump the gap between neurons.

TRUE / FALSE

4 Specific types of learning can affect the physical structure of the brain.

TRUE / FALSE

5 During sleep, activity in the cerebral cortex shuts down.

TRUE / FALSE

TEN THINGS A GENIUS KNOWS

1 The location of consciousness
Although the idea of the brain as the seat of mind and consciousness seems obvious these days, it wasn't always the case; the ancient Egyptians, for example, assumed that the heart did the thinking and therefore left it intact during mummification while extracting the brain and other organs. Later Greek philosophers differed –Hippocrates favoured the brain as the thinking organ, while Aristotle was a heart man, arguing that the brain was just a means of cooling the blood. Fortunately, the Graeco-Roman Galen, whose works had a lasting influence on medicine through the medieval period and beyond, at least worked out that the brain was the location where muscles were controlled and sensory signals processed. But it was not until the 16th century that the great anatomist Andreas Vesalius made the first useful maps of brain structure.

2 Origins of neuroscience
Many historians consider the true beginning of neuroscience to be Italian physician Luigi Galvani's 1780 discovery that he could use electricity to stimulate nerves in dissected frogs' legs. Through the 19th century, scientists made steady progress on identifying the functionality of different areas of the brain through a mixture of animal experimentation and observations of patients with damaged or underdeveloped brains. The ability to view small details of the brain for the first time through improved microscopy led eventually to Spanish pathologist Santiago Ramón y Cajal's proposal of the 'neuron doctrine', the modern view that the brain is composed of large numbers of separate neurons. Today's neuroscientists may use more sophisticated methods to image brain structure and activity, but they still work on the assumption that the brain is powered by many neurons working together.

3 Brain structure
The brain has several distinct regions that form a clear hierarchy in both structure and function. Lower reaches are generally in charge of basic functions common to most vertebrate animals, while the upper layers do the actual thinking. The lower parts (sometimes known as the 'limbic system') are in control of essential bodily housekeeping: they include the medulla oblongata (which controls breathing and heart rate), the cerebellum (muscle control), the hippocampus (crucial to memory) and the midbrain (temperature regulation, alertness, and much of the basic sensory hardware). The hypothalamus, meanwhile, pumps out a variety of hormones (chemical messengers) in conjunction with other glands in the body. It's closely linked to the thalamus, the heart of the limbic system where many of our most basic emotions, appetites and drives are generated.

4 Grey and white matter
The large upper portion of the brain, with its familiar folds and lobes, is known as the cerebrum. It's here that we do most of our actual conscious thinking, from the interpretation of sensory signals to the retrieval of memories and everything from daydreaming to long division. Its outermost layer, densely packed with neurons and forming the archetypal 'grey matter', is known as the cerebral cortex. It's covered in ridges (gyri) and grooves (sulci) that increase its surface area to pack in more neurons. Cortical neurons linked to 'executive functions' such as perception, memory and decision-making are, in turn, linked to white matter beneath, which relays signals to and from the lower brain (for instance, when conscious actions are required).

5 Hemispheres and lobes
Famously, the cerebrum is divided into distinct left and right hemispheres. Deep sulci divide each hemisphere into four lobes: the frontal, temporal, parietal and occipital (front, side, top and back respectively). Maps of brain activity suggest that, very broadly speaking, the frontal lobe is in charge of movement and executive functions, the temporal lobe handles hearing, the parietal lobe spatial awareness and the occipital lobe vision. In addition to its role in hearing, the temporal lobes seem to be involved in processing sensory information and triggering associated memories – a key aspect of interpreting the world.

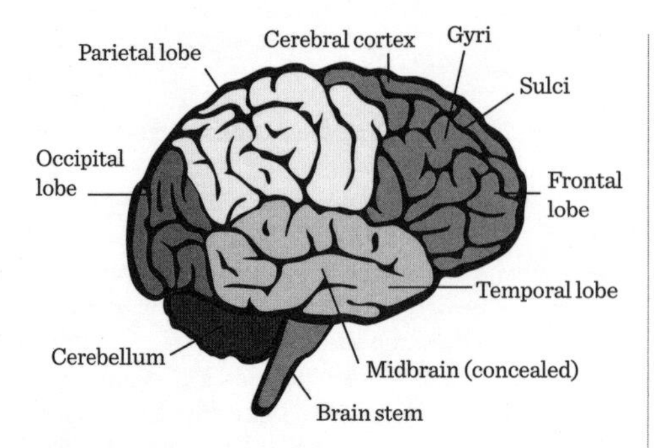

6 The left/right division

The popular division between logical left-brained and creative right-brained people is a gross oversimplification (we all use all the different parts of our brains at different times), but it has at least some grounding in reality, since each hemisphere does seem to have its specializations. It's well known that each hemisphere of the brain controls muscles and processes nerve signals from the opposite side of the body, but the left hemisphere also seems to do most of the work in language processing, logic and detailed calculation, while the right takes the burden when it comes to visual interpretation and spatial awareness. Normally, the outcome of this processing work ends up being shared between hemispheres via a bridging area called the 'corpus callosum'.

7 Neurons

The human brain is, at its heart, a collection of nerve cells, or neurons, broadly similar to those that form a network throughout our bodies. Each neuron is a specialized cell whose membrane has numerous rootlike extensions called 'dendrites', and a single long signal-carrying branch called the 'axon', which splits into multiple 'terminals' at its far end. Nerve signals travel along the axon as a so-called 'action potential' – a flow of electrically charged chemicals that is more like the separation of current inside a battery than the traditional image of an electrical spark. A gap called the 'synapse' separates the axon terminals from neighbouring dendrites.

8 Making neural connections

Each neuron in the brain may connect to hundreds of its neighbours, or just a few, and neuroscientists believe that 'plasticity', the ability to rearrange connections and form new ones, is the key to human learning. Synapses can be bridged either by the release of chemicals called 'neurotransmitters', or by an electric field generated by concentrated charge in the terminal influencing the dendrite (still no sparks, though!). Scientists spend much time mapping how networks of neurons in different parts of the brain 'fire' in response to various stimuli and during different types of brain activity. The essential approach of modern neuroscience can be summed up by the assumption that 'cells that fire together, wire together'.

9 Learning and memory

Perhaps the biggest challenge for science is to discover how the brain stores and retrieves memories, and how it puts these to use in learning. Brain scans suggest that storage and retrieval take place over large areas of the cerebral cortex, and that memory involves rearranging the network of connections across a large web of neurons. Short-term memory storage involves simply firing neuron signals around existing networks, but long-term storage is associated with the production of proteins that help to strengthen links between specific neurons. The hippocampi (one for each hemisphere) play a key role in transferring short-term memories to long-term storage, and also control the more instinctive spatial memory used in muscle coordination.

10 Controlling the body

The human nervous system is broadly divided into a 'central' nervous system (the brain and spinal cord) and a 'peripheral' system that extends throughout the body. In general, nerve signals travel from brain to body and back via the spinal cord, but some of the skull's cranial nerves (such as those from sense organs) have more direct routes to the brain. The peripheral system is further split into an 'autonomic' (involuntary) system and a 'somatic' (voluntary) system, and the autonomic system is always in one of two states: 'sympathetic' (which kicks the body's organs into 'fight or flight' mode) or 'parasympathetic' (which puts them in a more relaxed 'rest and digest' mode). In this way, the nervous system can flip the body rapidly from a state of relaxation to one of arousal.

TALK LIKE A GENIUS

' One of the most famous case studies in the history of neuroscience was a railroad worker called Phineas Gage, whose left frontal lobe was largely destroyed when a large iron bar went through his skull. Amazingly he survived, but his personality was completely changed and his behaviour became rude and vulgar. You could see the fact that he recovered a lot of his social graces in the 12 years before he died as early evidence for what we now call brain plasticity. '

' Some of the most unnerving neuroscience experiments involve patients who have had their corpus callosum cut as a way of reducing epilepsy. It turns out that if you sever that connection between left and right brain, then each hemisphere ends up with independent perception. Usually, that doesn't cause problems – but if you put a patch over the right eye, the patient won't be able to describe anything you show them. That's because the left eye routes to the *right* side of the brain, and the speech centre is usually in the left hemisphere. You can get even weirder effects by giving the two brains different types of information at the same time. '

WERE YOU A GENIUS?

1 FALSE – in fact, the brain uses up 10 times more energy per kilogram than the rest of your body.

2 FALSE – while those numbers are roughly correct, the distribution of connections varies hugely.

3 FALSE – there are no sparks in the brain; instead, electrical signals are carried by various chemical means.

4 TRUE – a study of trainee taxi drivers, for example, found that grey matter in the hippocampus region increased as they learned routes.

5 FALSE – the brain blocks conscious signals reaching the cortex, but the region itself remains active while we sleep.

Our unique brains combine vast networks of neurons capable of handling the higher functions that make us human, with ancient inherited structures that handle specific tasks.

Psychology

'Every aspect of thought and emotion is rooted in brain structure and function, including many psychological disorders and, presumably, genius.'

STEVEN PINKER

Human behaviour is endlessly complex and surprising, so it's little surprise that approaches to psychology, the study of how the brain governs our behaviour and interactions with the world, are also hugely varied. Only professionals have time to get to grips with all this bewildering complexity, but any wannabe genius needs to be equipped with a grounding in what the different branches of psychology actually do, and a few handy insights into the peculiarities of human behaviour that arise from the way our brains operate.

Psychology doesn't just help you understand the thoughts of others – it can also help you know when you're fooling *yourself*.

1 MRI scanners can pinpoint the specific neurons that fire during different types of brain activity.

TRUE / FALSE

2 The broad division between left- and right-brained people does have some grounding in psychological studies.

TRUE / FALSE

3 Outcome bias is a cognitive error we make by assessing the quality of our decision making on the basis of its outcomes alone.

TRUE / FALSE

4 Balance theory argues that we like other peoples' opinions to reflect our opinions of them.

TRUE / FALSE

5 The representativeness heuristic explains why we leap to conclusions about people based on their appearance.

TRUE / FALSE

TEN THINGS A GENIUS KNOWS

1 Types of psychology

While many earlier religions and philosophical schools attempted to get to grips with the human condition, psychology only really emerged from philosophy as a scientific discipline in the 18th century. Today, it takes in a huge variety of approaches including 'behavioural' (how a certain stimulus provokes a particular response in the body), 'cognitive' (how thought processes take place in the mind) and 'biological' (what physically happens to the brain during different types of thinking). 'Clinical' psychology, meanwhile, aims to understand and treat psychological disorders. Social, linguistic, developmental – you name it, there's a branch of psychology that investigates it.

2 Neuropsychology

The 19th-century fad for phrenology, which attempted to link the shape of the cranium to specific personality traits, has echoes in modern-day neuropsychology's more detailed investigation of the way that psychological processes and traits are linked to the physical and biological structure of the brain. At its heart lies the principle of 'modularity' – the idea that specific areas of the brain are specialized to carry out particular functions. This certainly seems to be true of functions such as sensory input, muscle control and metabolic regulation, which can be traced to specific localized regions of the brain. But the higher functions of conscious thought seem to require a different kind of modularity – the identification of networks across large areas of the brain that nevertheless seem to be associated with specific mental processes. Experiments conducted using modern brain-imaging techniques are vital when it comes to identifying these broader patterns.

3 Evolutionary psychology

Another field of psychology attempts to explain our mental quirks in terms of the various environmental 'selection pressures' our ancestors faced during their evolution. Its advocates argue that behavioural traits offering a positive benefit, in terms of general survival and/or successful mating and child rearing, will tend to propagate through the population. Disadvantageous ones, meanwhile, will tend to disappear over generations, in just the same way that physical attributes do. Evolutionary psychology can therefore offer a useful foundation for other, more specialized, psychological fields. It also has the advantage that (in contrast to some other branches of psychology) it puts forward testable theories and predictions about the ways in which people will respond in different situations.

4 Behavioural psychology

The behavioural approach to psychology assumes that nurture is far more important than nature, and so while inheritance has some role in our behaviour, we are mostly shaped by our experiences. We learn in infancy and throughout life to associate certain different stimuli with positive and negative outcomes, and our thought processes are therefore usually a balancing of these reward and risk possibilities. Reinforcement of particular stimuli, meanwhile (whether positive or negative) can produce gut reactions rather like physical reflexes. Behaviourism seems to exert a particular fascination for those studying psychology within organizations and cultures. Approaches vary between those who sift and analyse the results of large-scale surveys, and those who have a more qualitative approach based on close questioning and observation of small focus groups or anthropological field studies.

5 Cognitive psychology

The cognitive approach concentrates on understanding the actual processes our brains use to fulfil various tasks, rather than worrying about *why* we use those processes or where they came from. It covers topics as diverse as the processing of sensory signals, memory tasks, recognition and the formation of judgements based on available evidence. Most cognitive scientists believe that many of our day-to-day activities run on a sort of autopilot, while more demanding tasks are handled by 'executive functions'. According to one popular model, the brain breaks down cognitive tasks into subroutines called 'schemas', which are ordered and directed by either 'contention scheduling' (for the routine stuff) or by the 'supervisory attentional system' for more complex tasks. A major breakthrough came with the realization that many schemas produce quick results using intuitive short-cuts or rules of thumb called 'heuristics', rather than by processing problems from scratch.

6 Cognitive biases

The discovery of heuristics has revealed some intriguing biases that are hardwired into the thought process; they help explain why we sometimes instantly leap to wrong judgements, and also why we stick to them so rigidly. To give a few examples, the 'availability' heuristic measures the probability of events by how easily we recall examples; the 'representativeness' heuristic skews our judgement about the likelihood of events, depending on how well they match stereotypes; and the 'familiarity' heuristic is an assumption that behaviours that worked in the past will hold true in broadly similar situations, even when some key conditions have changed. The oft-quoted 'Dunning-Kruger effect', meanwhile, is the tendency for people with a little knowledge of a subject to assume they're experts (prospective geniuses beware!).

7 Our storytelling minds

Another commonly used mental heuristic is our tendency to create associations of cause and effect, even when they don't actually exist. A false association between disconnected events lies at the heart of much superstition. It's also the root of the so-called 'gambler's fallacy', a cognitive bias that wrongly assumes the chance of a coin landing heads-up on the next throw is increased the more times it lands heads-down. Despite the trouble this story-building tendency sometimes gets us into, it may also be the price we pay for the ability to visualize the consequences of our actions and attempt new behaviours as a species.

8 Social psychology

Very few of us live in isolation, and psychology is at its most revealing when it attempts to understand our social interactions. Social psychologists use various tools to understand how our attitudes, behaviour and even personalities are shaped by the influence of others, including public opinion surveys, detailed case studies and controlled experiments. One famous aspect of social psychology is the 'halo effect', identified by Edward Thorndike as early as 1920. This cognitive bias is the tendency for a good or bad impression of one aspect of a person or thing (typically, their appearance) to affect our perception of their other qualities and decisions. For obvious reasons, the advertising industry makes great use of this particular insight.

9 Balance theory

One central idea of social psychology is 'balance theory', formulated by Fritz Heider in 1958. Heider argued that our brains don't like cognitive inconsistency, so we tend to expect people to share our likes and dislikes (positive or negative 'valences' to use the jargon). Balance is achieved between individuals if they like each other (or even share a mutual dislike), but one-sided relationships are psychologically unsettling. Less obviously, we find it comforting if the attitudes of other individuals towards third parties, concepts or objects tally with our view of the individual (so we want people we like to like the same things we do, but if we learn that someone we dislike shares our passions, we may warm to them).

10 Applied psychology

Aside from using psychology to learn about the mind's inner workings, we can also use its insights to improve our thought processes, and overcome mental health problems, such as depression and phobias. Medical treatments use drugs to treat imbalances in brain chemistry, while other therapies attempt to identify problems through talking, or treat them through mental exercises to reorganize the brain. The different approaches taken by various talking therapies are sometimes controversial and the evidence for them varies. The psychoanalysis of unconscious desires, pioneered by Sigmund Freud in the late 19th century, has suffered a marked decline in standing in recent decades, with cognitive behavioural therapy, an approach that involves 'coping strategies' to suppress a patient's problematic thinking habits, rising in its place because of much stronger evidence that it actually works.

TALK LIKE A GENIUS

‘ One of the big barriers to neuropsychology is the resolution of current imaging techniques. MRI can pin down active areas of the brain to about 3 mm (⅛ in), but that area could still contain millions of neurons, so it's hard to pin down specific activities to particular pathways. ’

‘ Want an easy example of the availability heuristic? In just 10 seconds, write down as many words as you can think of starting with the letter K. Now write down words with K as the third letter, again in 10 seconds. I'm guessing you've got less on the second list, yet there are about three times as many English words in that group – you just can't think of them as easily. It's the same kind of mistake that makes people more scared of terrorism than they are of crossing a busy road. ’

‘ Some critics go even further with the idea of storytelling as a product of evolutionary psychology, arguing that the basic templates of mythology like the "hero's journey" (think Perseus and Luke Skywalker) developed as a means of keeping the tribe together and preventing unruly youths from getting out of hand. ’

WERE YOU A GENIUS?

1 FALSE – no brain scanner can yet show better than small regions of activity.

2 FALSE – although some brain functions are specific to each side of the brain, elements such as learning, creativity and problem-solving seem to be broadly distributed across both hemispheres.

3 TRUE – this can lead us into overconfidence when situations with a large element of chance happen to work out for the best.

4 TRUE – in brief, we like people we like to like things that we like.

5 TRUE – we tend to make assumptions about people based on how well they match our off-the-shelf stereotypes.

Together, the varied aspects of psychology reveal the complex story of how we think, but perhaps their most important insight is to show how we make mistakes.

The hard problem

'We know nothing about the intrinsic quality of physical events, except when these are mental events that we directly experience.'

BERTRAND RUSSELL

The so-called 'hard problem of consciousness' is the question of how our brains assign properties to certain sensory experiences; how, for instance, do we process phenomena such as colour or taste? Philosopher David Chalmers coined the phrase as recently as 1995 (contrasting it with so-called 'easy problems' like memory storage and instinctive reactions), and it's since become a hot topic for discussion. Some psychologists find the hard problem's philosophical aspects disconcertingly woolly, but there's no denying that it has deep roots in Western thought.

The hard problem is where neuroscience and philosophy meet – and there are many different ideas about how to bridge the gap.

1 The traditional mind-body problem is concerned with the ways in which our immaterial minds control our physical bodies.

TRUE / FALSE

2 Isaac Newton's work on colours of light led him to an early recognition of what we now call the hard problem.

TRUE / FALSE

3 The hard problem is generally seen as a challenge to the purely physical origins of the mind suggested by neuroscience.

TRUE / FALSE

4 So-called 'neural correlates' are the responses of individual neurons to specific stimuli.

TRUE / FALSE

5 Quantum consciousness is the idea that higher consciousness evolves from quantum-mechanical interactions between neurons.

TRUE / FALSE

TEN THINGS A GENIUS KNOWS

1 The mind-body problem

The idea of a distinction between mind and body is deeply rooted in the way we humans interpret the world, perhaps for fundamental evolutionary reasons (see page 29). Most religions take it as read, with the physical body merely a container for an immaterial and possibly immortal soul, but 'dualism' really took root as a concept for natural philosophers in the writings of René Descartes. In particular, his argument that thinking is the only thing that makes us sure of our existence ('I think, therefore I am') privileged mental activity and suggested that, even if robbed of our senses or deluded by false impressions, we would still have 'mind'. The question of how a material body interacts with an apparently immaterial mind, and mind commands a physical body, however, is a fundamental one that continues to intrigue scientists and philosophers, even as we've learned more about the brain as an organ.

2 Dualism and monism

From a philosophical point of view, the two major approaches to the problem as to how brain and body interact are 'dualism' (broadly following the ideas of Descartes) and 'monism', the idea that body and mind are, despite our perception, manifestations of a single substance. Dualist solutions to the mind-body problem include those of Gottfried Leibniz (who got around the problem by suggesting that mind and body don't actually interact, but just appear to because both are programmed to follow a pre-established harmony). Monist ideas range from the materialism of Thomas Hobbes (who argued that mind is effectively a creation of matter – essentially, the view of most modern science) to the 'neutral monism' of Immanuel Kant (who suggested a single substance with intermediate properties, capable of somehow manifesting as both matter and mind).

3 The conscious Universe

During the 19th century, a popular alternative model of mind known as 'panpsychism' considered consciousness to be a property of the Universe itself – in other words, mind is an innate property of all things and we differ in our level of consciousness from other animals, plants and even inanimate objects only by degree. Alfred North Whitehead had his own take on this idea, arguing that 'occasions of experience' are the fundamental elements of reality and that these can accumulate to create complex entities, including ourselves. This was the core of Whitehead's 'process philosophy', an influential approach that considers change and experience, rather than objective reality, to be the best explanation for how we understand the world.

4 Medical solutions

From a purely scientific point of view, however, medical breakthroughs in the 19th century settled the immediate mind-body question once and for all. The development of anaesthetics from the 1840s onwards, showed that physical intervention could render a patient unconscious, effectively confirming that mind must be a manifestation of body. (The precise mechanism of general anaesthetics is still poorly understood, though they're generally agreed to block the transmission of signals across the synapses between neurons.) Attention now began to turn to a new form of philosophical question – how do signals from the outside world create impressions that manifest in what we call mind?

5 Qualia

Neuroscientists and philosophers alike use the term 'qualia' to describe our conscious perceptions of different sensory experiences. The word, derived from Greek and meaning roughly 'of what type', was coined by American philosopher Clarence Irving Lewis in 1929. Formed in the mind, qualia are subjective – for instance, phenomena such as red-green colourblindness show that two individuals can have quite different perceptions when their retinas are illuminated by different wavelengths of visible light. And qualia are also frustratingly hard to describe directly to other people without using language that implies a common understanding in the first place. Philosopher Thomas Nagel famously highlighted this problem of common ground in a 1974 essay entitled 'What is it like to be a bat?' In brief, Nagel argues that, while we can begin to imagine what it might be like in our own terms, we have no way of knowing how the bat experiences the world.

6 Philosophical zombies

The fact that qualia are ultimately subjective can lead to an interesting philosophical argument – although thought itself is enough to prove that 'I' am a conscious being, what about everybody else? Could they just be faking it? It seems perverse to suggest that everybody else is a 'philosophical zombie' producing outward signs of emotion and engagement but no actual inner consciousness (and common sense, of course, argues otherwise), but the argument does at least help to point out that we are flawed in believing there is inescapable logical evidence to support the consciousness of others.

7 Neural correlates

Breakthroughs in imaging brain activity have led many neuroscientists to study the so-called 'neural correlates of consciousness' (NCCs). These are the smallest possible patterns of brain activity associated with specific 'percepts' (stimuli such as external sensory signals, memories or emotions). Artificially stimulating NCCs can trigger specific percepts, while deactivating them can cause percepts to disappear (and various tricks can be used to prevent a physical stimulus in the outside world from triggering the expected percept in the brain). The main lesson from these studies so far is that the brain is 'plastic' – networks of neurons can rewire themselves in response to traumatic damage or simply as we learn new skills, so that the neurons associated with particular percepts are not always the same from subject to subject and/or over time.

8 Stating the hard problem

Despite all the breakthroughs of neuroscience, however, some philosophers would argue that we're no closer to understanding how the brain bridges the gap between physical stimuli and mental experience or percept. This is what Australian philosopher David Chalmers described as the 'hard problem of consciousness' in 1995: 'Why is it, that when our cognitive systems engage in visual and auditory information-processing, we have visual or auditory experience: the quality of deep blue, the sensation of middle C? How can we explain why there is *something it is like* to entertain a mental image, or to experience an emotion?' In search of a solution, Chalmers argues for an irreducible universal property of consciousness reminiscent of panpsychism.

9 The claustrum mystery

In 2014, neuroscientists stumbled across a potentially crucial clue to the whole question of consciousness while investigating a region called the 'claustrum' (a thin sheet of neurons near the centre of the brain). During attempts to treat a woman with severe epilepsy using deep electrical probes, they found that the patient immediately lost consciousness when a current was applied in this region, and recovered with no intervening memory when the current was removed. The claustrum seems to play a role in the collation of sensory information, and some neuroscientists have argued that it may be central to coordinating information from different parts of the brain and giving rise to the phenomenon of consciousness. If the hard problem has a biological solution, then, it may be found here – but there is much more investigation to be done.

10 Quantum consciousness

One intriguing sideline to the hard problem is the controversial suggestion that consciousness could arise from quantum-mechanical processes that occur in tiny brain structures called 'microtubules'. Proposed by Oxford mathematician Roger Penrose and psychologist Stuart Hameroff, the 'Orch-OR' theory argues that mental processes cannot be described by even the most complex algorithm. It then suggests that unique mental capabilities, such as creativity and problem solving, arise because our brains are in fact quantum computers, with qubits of data (see page 178) processed inside microtubules and amplified by the overall neuron structure. The Orch-OR theory has been criticized for its suggestion that consciousness arises within neurons, rather than via connections between them, and some have questioned how qubits could stay in an uncertain quantum state in such a 'warm, wet and noisy' environment. In 2014, however, scientists in Japan confirmed that quantum vibrations do indeed take place inside microtubules, so the debate looks set to continue.

TALK LIKE A GENIUS

“Cognitive scientist Daniel Dennett once said qualia are best understood as simply “the way things seem to us”, and that description hits the nail on the head. How do we know how the same things seem to others? Try describing a smell without lapsing into metaphor, or even think about a colour; how can you know that just because you and a friend both agree that a balloon is “red”, you're both perceiving exactly the same thing?”

“Of course not everyone accepts that there is a “hard problem” to worry about in the first place. Some people just say there is no such thing as a separate consciousness, just a piling up of experiences that can be recalled from the mind in response to various stimuli, while others say that qualia don't really exist in the first place. Perhaps the most famous put-down of the whole idea came long before David Chalmers had even dreamt up the modern hard problem, when back in 1949 Gilbert Ryle decried the whole idea of dualism as a “category mistake”, a bit like looking for the ghost driving a machine.”

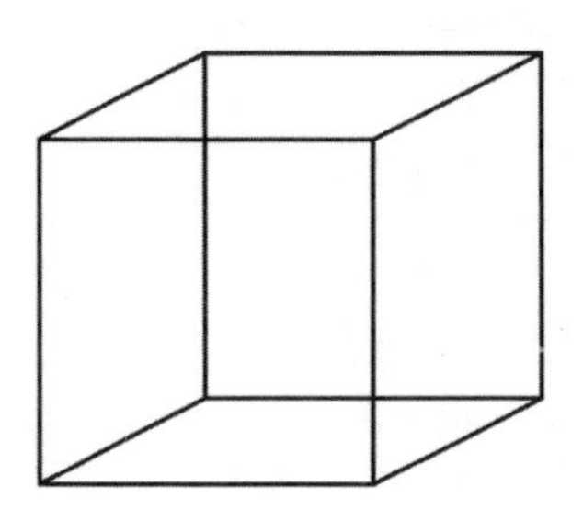

The ‘Necker cube’ illusion can be considered as a single sensory input giving rise to two distinct qualia in our minds.

WERE YOU A GENIUS?

1 TRUE – the mind-body problem and the hard problem both address the gap between the physical and mental, from different directions.

2 TRUE – as early as 1672, Newton questioned the process that causes us to perceive different types of light as different colours.

3 TRUE – or at least this was Chalmers' intention, though many neuroscientists deny there's a hard problem at all.

4 FALSE – neural correlates mostly involve large networks of neurons rather than individuals.

5 FALSE – the model suggests that consciousness is created *inside* the individual neurons.

How and why does the brain assign qualities to sensory stimuli? Possible solutions to the hard problem depend on whether or not you think consciousness is entirely physical.

Artificial intelligence

'May not machines carry out something which ought to be described as thinking but which is very different from what a man does?'

ALAN TURING

Can a machine think? It's a question that still feels like the stuff of science fiction, but recent advances in the field of artificial intelligence (AI) are increasingly allowing computers to do tasks that were once thought to have uniquely human abilities. The rise of artificial intelligence raises profound challenges across fields ranging from economics to ethics, but while some see it as a disruptive challenge to current society, or even an existential threat, it also carries a huge promise for the future of humanity.

We're getting closer to artificial intelligence all the time, but should we be careful what we wish for?

1 The Mechanical Turk was a famous 18th-century automaton capable of beating most people at chess.

TRUE / FALSE

2 Neural networks learn complex tasks by a form of machine evolution that allows them to improve over time.

TRUE / FALSE

3 After being beaten by the Deep Blue AI in 1997, chess champion Garry Kasparov accused his opponent of cheating.

TRUE / FALSE

4 A Turing machine is a device that is capable of passing the Turing Test for artificial intelligence.

TRUE / FALSE

5 Most AI advocates agree that artificial intelligences will require some inbuilt ethical programming to prevent them from harming humans.

TRUE / FALSE

TEN THINGS A GENIUS KNOWS

1 Origins of AI

The dream of creating a thinking machine goes back to antiquity – artificial beings with some ability to think for themselves feature in legends of the Greek blacksmith god Hephaestus, and, of course, the story of Pygmalion and his animated statue. Medieval philosophers continued the obsession, as attested by folktales of the Golem conjured by Rabbi Judah of Prague, the *Takwin* pursued by Muslim alchemist Jabir ibn Hayyan, and the talking brass head supposedly built by Roger Bacon. Alongside these fantasies, philosophers were doing the more serious work developing systems of logic and reason. Majorcan Ramon Llull (*c.* 1232–1316), who built paper machines to find 'truths' using a symbolic alphabet, was a significant influence on Leibniz and therefore on the whole field of symbolic logic and computation. By the early 19th century, a growing understanding of the interaction between brain and body inspired Mary Shelley to create perhaps the most famous 'artificial intelligence' in all of fiction: Frankenstein's monster.

2 The Turing Test

English polymath Charles Babbage is famous for his mechanical difference engine, an early calculator invented in 1822, and for later realizing that a far more versatile 'analytical engine' was possible. Modern computers, however, are electronic and digital (they treat data as strings of binary 1s and 0s in order to speed their calculations). The famous code-breaking Colossus, built in secret by the British during World War II, was the first programmable digital computer. Apart from pioneering mathematician Alan Turing, however, few saw that this versatile machine might one day be capable of mimicking a human brain. In 1950, Turing proposed his 'Turing test' for AI; it involves a human judge conversing with both a computer and a real human via printed text; Turing argued that a machine could reasonably be said to display intelligence if the judge could not tell which of their conversation partners was which.

3 Neat and scruffy AI

The first conference on 'thinking machines', held at Dartmouth College, New Hampshire, in 1956, gave rise to some key ideas (including the term 'artificial intelligence' itself, coined by US scientist John McCarthy). Over the following decades, it became clear that achieving functional AI would be much harder than planned. Early research involved two broad approaches: so-called 'neat AI' attempted to develop a model of intelligence using standard programming techniques, while 'scruffy AI' involved bolting together different elements in a more haphazard way, in the hope of discovering 'emergent' behaviours that mimicked human intelligence.

4 Expert systems

One 'neat' approach to AI involves teaching a computer everything it needs to know about a particular task or field, and the way that human experts make decisions in that field. Beginning in the 1970s, so-called 'expert systems' were heralded as the first successful application of AI. Typically, they consist of a 'knowledge base' outlining all the relevant facts and rules of a particular field, and an 'inference engine' that interprets questions or tasks by applying logical procedures to those rules and data. The development of expert systems coincided with the rise of personal computers, and for a while they became hugely popular. Indeed, their basic principles are still widely used, even if the terminology has fallen out of fashion.

5 Neural networks

From the 1980s, a more successful and versatile form of AI emerged from the 'scruffy' approach. Neural networks are arrays of relatively simple software 'machines' that share data and form connections in similar ways to the neurons in our brains, with signals propagating through multiple layers of neurons from an input layer to an output layer. These signals can be as simple as a digital '0' or '1', or more complex with varying strengths, and thresholds or limits to the triggering of an output signal. Such mimicry of the human brain structure allows neural networks to develop problem-solving procedures based on exposure to examples – so-called 'machine learning' that avoids the need to explicitly program complex tasks.

6 Recognition tasks

Neural networks are particularly suited to challenging tasks such as speech and handwriting recognition and more general computer vision. Ask half a dozen people to write down the digits 0 to 9 and you'll see how variation might make it difficult for a computer to identify them from strict rule-based programming. Yet our brains do this same task intuitively and if you give a neural network a large set of examples and some basic processes, it too will form its own connections and find out its own procedure to rapidly identify new numbers. The clever (and perhaps disconcerting) bit is that we don't need to worry about what's going on in the intermediate 'hidden' layers of the network – we only have to concern ourselves with the input and output layers.

7 Human vs machine

Computing firm IBM grabbed the limelight in 1997 when its chess-playing computer Deep Blue narrowly won a tournament against world champion Garry Kasparov (the machine used smart algorithms to follow the most 'interesting' available moves to their possible conclusions, and carried a reference library of possible endgames). In 2011, another IBM system, Watson, beat former champions on US quiz show *Jeopardy!* despite problems interpreting the grammar and context of questions. Once it had parsed the question into a more computer-friendly format, Watson used a sophisticated system to search its onboard reference library and rank possible solutions. Since 2015, AlphaGo (a system based on Google's DeepMind machine learning system) has beaten a series of champions in the ancient Chinese board game Go, widely considered to be even more complex than chess.

8 What AI can do

Since the 1990s, meanwhile, AI has become increasingly prevalent in the real world. Early applications included Sony's robotic dog Aibo, which had a limited capacity for learning techniques and developing a distinctive 'personality'. Somewhat more mundane are robotic vacuum cleaners and lawnmowers that can learn the layout of rooms and respond to unexpected obstacles. Driverless cars, meanwhile, address similar but vastly more complex challenges where split-second timing is required and lives are at risk; they are already reaching a point where they are statistically safer than human drivers, and unlike humans have the potential to rapidly improve even from a high benchmark. Whether the public are willing to put their lives directly in robotic hands may be another matter, however.

9 The singularity

Ever-improving miniaturization techniques and the young, but promising, field of quantum computing (see page 178) suggest that computers will continue to get faster and more powerful, at least for the foreseeable future. Coupled with improved AI techniques, many computer scientists and futurologists predict the eventual emergence of AIs that are capable of designing a new generation of AI better than themselves. At this point, a process of exponential improvement will begin, rapidly culminating in an AI of effectively limitless intelligence that mathematician John von Neumann called a 'singularity'. Futurologist Ray Kurzweil predicts that the singularity will arrive before the end of the 21st century, but many others are not so sure, predicting that technological barriers or inherent properties of mind may render the singularity unreachable.

10 Conscious computers

Related but distinct from the singularity is the question of whether computers can ever achieve true consciousness, or whether they will just get better and better at faking it. Neuroscientists and philosophers still wrestle with the so-called 'hard problem' of bridging the gap between physical stimuli and mental qualia in organic brains, and understanding how an electronic brain could make a similar leap to human-like consciousness might seem a far bigger challenge. But if the hard problem can be solved, then its resolution may reveal how to create truly brainlike machines capable of bridging the gap. Nevertheless, some theorists believe that consciousness is far more than 'just' the hard problem; it's also the web of connections we intuitively create between concepts and experiences, and *that* may prove far harder for a computer to mimic.

TALK LIKE A GENIUS

❛ You know those CAPTCHA letter-identification tests you sometimes have to fill in before submitting an internet form? Well CAPTCHA stands for Completely Automated Public Turing test to tell Computers and Humans Apart. It's actually a sort of reverse Turing test because in this case there's a computer on the other end of the form that's been told the code already and is testing whether you can use your human pattern recognition skills to read the picture correctly. ❜

❛ Computer scientists have wanted to beat the traditional Turing test since the 1960s, and have spent a great deal of time and effort programming "chatbots" aimed at mimicking human conversation. In 2014, a Russian chatbot called "Eugene Goostman" got a lot of publicity for apparently beating the test, but since the program ducked some of the questions by masquerading as a teenager with limited English, you could argue that it was bending the rules. ❜

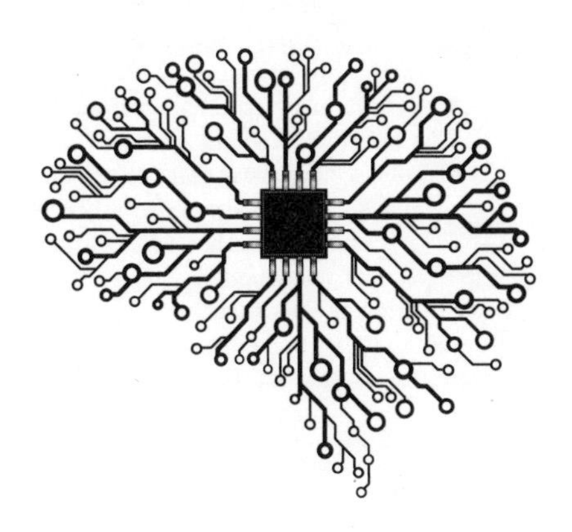

WERE YOU A GENIUS?

1 FALSE – the Turk was in fact a sophisticated puppet operated by a concealed human chess player.

2 TRUE – feedback mechanisms allow networks to judge their performance and try new approaches.

3 TRUE – Kasparov suspected human intervention and requested that IBM release details of the machine's operation. IBM refused and dismantled the computer, despite Kasparov's request for a rematch.

4 FALSE – a Turing machine is a computer defined by its ability to perform various mathematical operations.

5 TRUE – as early as the 1940s, sci-fi author Isaac Asimov imagined three laws of robotics controlling AI behaviour in a future society. Today, his laws still frame much of the debate around AI ethics.

Computers can fake intelligence in increasingly convincing ways. There are doubts about whether they could become conscious, but perhaps they can fake that too.

History of art

'The aim of art is to represent not the outward appearance of things, but their inward significance.'

ARISTOTLE

Art is one of the oldest human endeavours, going back to at least 35,000 BCE and intrinsically linked, it seems, to the origins of human consciousness. As such, artistic representation is one of the main ways (and sometimes the only way) for us to interpret the civilizations and cultures of the past. It's the sort of thing, therefore, that any genius should know about – and if you don't have time to get to grips with long lists of different artists and movements, then a few choice facts can take you a long way...

Finding an all-encompassing definition for art itself is an insurmountable challenge, but knowing *why*, *how* and *when* is a good start.

1 The definition of art as we understand it today emerged during the early 19th century.

TRUE / FALSE

2 A big reason for Renaissance paintings looking more realistic is that architect Filippo Brunelleschi invented the principles of perspective in 1413.

TRUE / FALSE

3 The Greeks and Romans preferred their statues to be made from pure white marble in order to distance their appearance from the everyday world.

TRUE / FALSE

4 The Virgin Mary is traditionally depicted in blue as a symbol of both her heavenly grace and her servitude to God.

TRUE / FALSE

5 The *Mona Lisa* was generally regarded as one of Leonardo's lesser works for most of its history, and only became the most famous painting in the world after it was stolen in 1911.

TRUE / FALSE

TEN THINGS A GENIUS KNOWS

1 What is art?

Art seems to be something innate to humans; from the dawn of consciousness we've felt driven to make objects and images that express both our inner feelings and our understanding of the world. While most art before the modern era is broadly representational of the real world, the artist's role in creating the finished work inevitably introduces layers of interpretation and meaning to even apparently straightforward imitations of life. Increasing recognition of this mediating role during the Romantic period ultimately gave rise to today's conception of art as a product of creative genius.

2 The first art

The earliest widely accepted works of art are sculptures made by early Europeans of the Aurignacian culture at the height of the last Ice Age. Aurignacian art includes exaggerated Venus figurines and eerie human/animal hybrids – perhaps protective talismans or tools for shamanic rituals, but don't believe anyone who says they know for sure. The oldest dated cave paintings – handprints and animal figures from Sulawesi, Indonesia – date to around the same time , but the famous European cave art (Lascaux, etc.) is the work of the so-called Magdalenian culture, about 15,000 years later.

3 Art and the ancients

While the remarkable animal sculptures found at Göbekli Tepe in Turkey (*c.* 9000 BCE) appear to be the work of an early hunter-gatherer society, it was not until agriculture took hold and enabled permanent settlement (and specialization of labour; no need for everyone to be foraging or hunting all the time) that art really took off. Stonework tend to outlast paintings and domestic objects, biasing the archaeological record so that, from Mesopotamia and Egypt to China and the classical Mediterranean, most surviving art tends to be sculpture associated with funerary rites, worship and imperial power. Rare domestic fragments, such as beautiful wall paintings from Pompeii, lost jewellery and discarded glassware, show that people also appreciated art in everyday life.

4 The religious imperative

The rise of Christianity in the Roman Empire from the fourth century CE led to a flourishing of art representing religious themes, ranging from exquisite Byzantine mosaics to stunning icons, reliquaries and devotional sculptures. Hindu art in India trod a similar path, while Buddhism developed its own tradition that headed East into China and beyond. The rise of Islam in Africa and the Middle East, meanwhile, had a contrasting effect, as the Muslim prohibition on human and animal representation gave rise to beautiful geometric, calligraphic and floral decoration schemes in grand mosques and palaces. Across all cultures, however, religious and secular authorities acted as sponsors for the creation of art intended to inspire awe and contemplation.

5 The Renaissance

Spanning roughly the 14th to the 16th centuries, the Renaissance (literally 'rebirth') was a period characterized by a newfound philosophical 'humanism' and the detailed study and improved expression of human anatomy in art. The term was first coined by painter Giorgio Vasari in his 1550 *The Lives of the Artists*, to define a period in Italian art beginning in the time of Giotto di Bondone (1266–1337) and culminating with Leonardo da Vinci (1452–1519) and Michelangelo (1475–1564). Influenced by new discoveries and reassessment of antiquities from the classical world, Renaissance thought was nevertheless marked by a shrugging-off of received opinion and a search for new approaches.

6 The Renaissance beyond Italy

Beyond the undoubted glories of Italy, equally impressive art was being created elsewhere in Europe during the Renaissance period. The expressive German wood sculpture of Tilman Riemenschneider and the realism of Dutch masters like Jan van Eyck and Pieter Bruegel the Elder rival anything the Italians have to offer. A new spirit of inquiry that accompanied the Protestant Reformation, along with cheaper transmission of ideas through the newly invented printing press, saw humanist ideas and new art practices spread rapidly across Europe. As Rome launched its own Counter-Reformation, Catholic art grew wilder and more extravagant (typified by the baroque), while puritanical Protestants often took the Bible's injunction on graven images to heart, indulging in bouts of iconoclastic destruction that robbed many religious buildings of their great medieval treasures.

7 Power and patronage

When looking at any piece of art, it's worth considering who commissioned it, and why. The Renaissance, for example, was fuelled by the rival egos of Italian nobility. When Lorenzo de' Medici sponsored religious altarpieces to ensure his eternal salvation, he made sure to be represented among the saints so his munificence wouldn't be forgotten on Earth. Reformation Holland might have shown a more subdued taste for 'genre' scenes of domestic life, but patrons still liked to showcase their finest possessions, ranging from cattle to tulips. More recently, while some industrialists and businessmen have commissioned art for the love of it, others have done so to prove their high-minded credentials and donated it to broadcast their commitment to society. Civic authorities, meanwhile, have backed art to commemorate the great and good, support ideologies, impart moral lessons, and sometimes just to excite debate and improve public spaces.

8 Art and technology

Art is inevitably shaped by the tools with which we make it. Paints, for instance, usually consist of pigments mixed in a suitable binding medium, but both availability of pigments and choice of binder have changed over time. The earliest paints were watercolours (China and Japan have long traditions of painting with ink, and the fluid, calligraphic brush strokes this permits are innately linked to their unique styles). Binders such as beeswax ('encaustic', giving a textural, even sculpted, finish) and egg yolk ('tempera', a faster-drying paint with a flat finish) were popular in Europe from Classical to medieval times, as was fresco (adding pigment to still-wet wall plaster). Fast-drying tempera and fresco, in particular, dictated rapid painting styles, but oil paints (invented in Afghanistan, but finally reaching Europe in the Renaissance) became popular for their longer drying times and greater control over layering, lending themselves to a more detailed approach. Brush types have their own influence on the art that is produced, as indeed do the various tools and media used in sculpture.

9 Post-Renaissance movements

The development of art from the Renaissance to the 19th century can be crudely characterized as a series of revivals and reactions. Sixteenth-century Mannerism developed late Renaissance ideas about proportion and beauty but eventually went over the top, resulting in stilted, hyperreal paintings. The 17th-century baroque, in contrast, was all about movement and drama, but it nevertheless retained strict rules of composition that its final flourish, the rococo period of the mid-1700s, kicked against with playful and even grotesque exuberance. About the same time, the far more austere approach of Neoclassicism naturally popped up as yet another reaction, harking back to ancient times and inspired by the discovery of Pompeii. Similarly, perhaps the last gasp of traditional art, the lush Romanticism of the 19th century can be seen as a reaction to both classical order and the rise of science and industry. Needless to say, all of these movements ranged far beyond painting, affecting everything from sculpture and architecture to literature and fashion.

10 Symbolisms

One final way of looking at art is as a series of symbols. Often symbolism is explicit, as in Renaissance depictions of mythological and religious scenes that are designed to leave the viewer with a certain moral message. Skulls, ornate jewels and females clutching apples carry connotations of impermanence, worldly wealth and temptation that are still with us today. But sometimes the symbolism is more cunningly concealed, or lost in the passage of time to a modern viewer, who may not know that the loose tresses of a Pre-Raphaelite beauty suggest prostitution, or that the countless flowers in Botticelli's *Primavera* carry an intricate Neoplatonist code. In the late 19th century, an entire artistic movement (Symbolism with a capital 'S') painted deeply complex works layered with somewhat self-important messages.

TALK LIKE A GENIUS

‘Archaeologists working in South Africa have found little sticks of ochre that were clearly scratched with crisscross patterns about 70,000 years ago – is that art? If we don't know the motivation behind it, how can we be sure?’

‘Art has always had a close relationship with technology, but it may have been even closer than we generally assume. Some art historians, for instance, have argued that the newfound accuracy of Renaissance art is down to the use of optical devices like the *camera obscura* that projected images of reality onto the canvas to help guide the artist's hand.’

‘Graffiti is nothing new – Pompeii and other ruins are full of rude pictures and inscriptions, of course, but a really intriguing example are the *pittura infamante* (defaming portraits) from Renaissance Italy. If someone was crooked, unpopular and above the law, they might wake up one day to find a life-size picture of their own execution painted on a wall somewhere prominent. None of them survive today (and most of them probably got scrubbed off within a few hours unless they triggered a revolution) but we know from historical records that even famous artists like Botticelli got in on the act.’

WERE YOU A GENIUS?

1 TRUE – it was during this Romantic period that art, hitherto perceived as skill or mastery, was redefined as a unique faculty of the human mind.

2 TRUE – although many earlier artists had their own systems, Brunelleschi was the first person to work out the geometric rules for perspective as we generally understand it.

3 FALSE – although classical sculptors prized marble for its qualities as a working material, the finished articles were usually covered in colourful pigments or gilding.

4 TRUE – the blue of Mary's robes is part of a rich pattern of colour symbolism in medieval painting. Blue pigments were extremely costly, so they were reserved for depictions of heaven and important religious figures.

5 TRUE – the *Mona Lisa* only shot to real fame after it was stolen by Italian Vincenzo Peruggia in August 1911.

If you really want to understand a piece of art, you'll need to look beyond what's on the canvas.

Modern art

'The modern artist, it seems to me, is working and expressing an inner world – in other words ... expressing his feelings rather than illustrating.'

JACKSON POLLOCK

There are few subjects as divisive as modern art, and perhaps that's as it should be, since, as the quote above explains so well, modern art's primary function is to engage with the inner thoughts and feelings of artist and viewer. The wide variety of artistic movements of the past 150 years can sometimes be confusing, however, and if you want to venture an opinion, it's handy to at least know what the artist was trying to do.

Modern art might not be pretty, but the question is what does it make you feel or think?

1 The name impressionism comes from an off-the-cuff remark by Monet.

TRUE / FALSE

2 A black square painted by Kazimir Malevich in 1915 is sometimes regarded as the most influential work of abstract art.

TRUE / FALSE

3 Early cubists were influenced by developments in maths and physics during the early 20th century.

TRUE / FALSE

4 The first formal recognition of graffiti by the art world was an exhibition at the Brooklyn Museum in 2006.

TRUE / FALSE

5 Outsider artists are a school that deliberately distance themselves from the art establishment.

TRUE / FALSE

TEN THINGS A GENIUS KNOWS

1 What is modern art?

Ask a dozen experts, and you'll get a dozen different definitions of what modern art is and how, exactly, it took hold. One useful way of looking at it, however, is a process of accelerating innovation and reinvention. The deep history of art is certainly scattered with innovators capable of presenting startlingly modern and non-naturalistic visions. However, the invention of photography in the mid-19th century *forced* artists to look beyond direct visual representation to think more deeply about the thoughts, feelings and reactions their work might inspire. Arguably, they never looked back.

2 Japonism

The arrival of Japanese influence in Europe is often seen as ground zero for the modern art revolution. Japan had only reopened to trade with the West in 1853, and stylized arts and crafts developed during Japan's isolationist period (think Kakiemon porcelain and the block prints of Hokusai, famous for his *Great Wave*) began to reach Europe soon after. Here, artists such as Tissot, Degas, Whistler, Monet and Manet began to use them as inspiration for their own work. The movement gathered strength after the first formal exhibition of Japanese art at the 1867 Paris World's Fair, and was named 'Japonism' by French critic Philippe Burty a few years later. Its many forms inspired textile and porcelain designs, European-style depictions of Japanese subjects but, more importantly, also radically new ways of presenting art itself (for instance, the mass-produced posters of Toulouse-Lautrec, and the stylized lines of art nouveau design).

3 Impressionism and post-Impressionism

The first major modern art movement, impressionism, was based on the idea of painting what the artist saw rather than an idealized version. Originating from the new habit of rapid outdoor painting known as '*en plein air*', impressionism typically involved applying small strokes of pure colour to rapidly build up a vivid depiction of subjects as they appeared under changing natural light. In keeping with this rapidity, subjects became more varied and less formally arranged, while movement also became a significant element. Major names include Pissarro and Sisley, Renoir and Degas. Post-Impressionism developed as a trend towards more abstract and symbolic images in the late 19th century and was inspired by new scientific breakthroughs in the understanding of colour and vision. It includes the complex and bold compositions of Cézanne, the vivid colours of Van Gogh and the pointillism (paintings composed of countless tiny paint dots) pioneered by Seurat.

4 Expressionism

In contrast to impressionism, the expressionist movement aimed to create subjective emotional effects, rather than depict the objective world. It developed in early 20th-century Germany as part of the more general modernism, taking inspiration from the introspection of Freudian psychoanalysis, Nietzsche's philosophy and gloomy Swedish playwright August Strindberg. In terms of art, its most obvious precursor is Edvard Munch's *The Scream* (1893), but it also borrowed heavily from traditional African art. Big names in pre-World-War-II impressionism include Kandinsky, Klee, Marc, Schiele and Kollwitz. Some claim the movement should encompass everything from cubism to surrealism, making it rather hard to pin down a particular visual style.

5 Modernism

Beginning around the start of the 20th century, the modernist approach was a loose movement united by a few shared attitudes, including an abandonment of historical rules, a tendency to abstraction and a belief in the transformative power of technology and design. Cubism is perhaps the most famous and influential modernist art school – pioneered by Picasso and Braque around 1907, and abandoning long-held principles of perspective and fixed viewpoints to create false illusions of depth. Italian futurism, meanwhile, created dynamic paintings and sculptures praising youth, power and machinery, while British vorticism did something similar but from a more angst-ridden viewpoint.

Functional, geometric constructivism flourished in post-revolutionary Russia from 1917, while the De Stijl group in the Netherlands (Mondrian and others) aimed to simplify art and design down to basic lines and minimal colours.

6 Duchamp invents conceptual art

The idea of conceptual art sprang out of the deliberately confrontational 'anti-art' movement that set out to challenge both artistic and political conventions in the midst of World War I. Marcel Duchamp rejected the idea that art should be intended to give visual stimulation, instead arguing that it should engage the mind itself. *Fountain*, a urinal with the scrawled signature 'R. Mutt', was submitted to a 1917 New York exhibition but rejected despite the exhibition's supposedly open-door policy. Duchamp flourished on the controversy, and produced several other 'readymades', of which perhaps the most famous is 'L.H.O.O.Q.', a cheap copy of the *Mona Lisa* adorned with a silly moustache.

7 Surrealism

Stemming from the same 'anti-art' ideas as conceptual art, Dada was a deliberately nonsensical blend of art, poetry and performance with followers in Zurich and New York. Its devotees protested against both the horrors of World War I, and the bourgeois society from which they had stemmed. Dada set the stage for the surrealism of the 1920s; a movement founded by poet André Breton that believed (taking cues from Freud) that the nonsensical juxtapositions of dreams offered deep insights into the subconscious mind. Salvador Dalí and René Magritte both created realistic imagery of dreamlike concepts and situations, while others such as Joan Miró and Max Ernst attempted to create disengaged 'automatist' art in an attempt to access their unconscious.

8 Abstraction

As the United States asserted its economic and cultural power after World War II, it naturally began to develop artistic movements of its own. Abstract expressionism was the first and among the best known, playing a key role in shifting the centre of the art world from Paris to New York. As its name suggests, artists involved abandoned traditional representations entirely (even the twisted ones of surrealism) in order to focus on expressing their innermost feelings in art. Jackson Pollock (1912–56) accomplished this by abandoning all attempt to control the placing of paint on the canvas, spraying it randomly as he moved in so-called 'action painting'. Willem de Kooning (1904–97) started off in this vein before returning to somewhat more controlled and representational pictures such as his *Woman* series. Others took more nuanced approaches, such as the 'colour fields' of Mark Rothko (1903–70), frameless blocks of painted colour that many nevertheless find deeply moving.

9 Pop art and after

The dominance of American popular culture gave rise to an artistic response in the form of pop art, which flourished in the 1950s and 1960s. Andy Warhol co-opted the techniques of advertising and traded on the growing fascination with celebrity to produce works, such as his famous Campbell's soup cans and portraits of Marilyn Monroe. Much of his art was mechanically reproduced by assistants in his famous New York 'Factory'. Roy Liechtenstein turned comic-book stylings into art, while British-born David Hockney produced bright and appealing domestic scenes influenced by the light and architecture of California. Back in Britain, meanwhile, Peter Blake's photographic collages (most famously the cover of The Beatles' 1967 *Sergeant Pepper* album) helped pave the way for later 'mixed media' art.

10 Postmodern art

Beginning with pop, much of the art of recent decades can be grouped together as 'postmodern', in the sense that it reacts to and challenges much of the modernist art that came before it. Postmodernism encompasses a bewildering range of artistic schools and styles – everything from the graffiti of Jean-Michel Basquiat and Banksy to performance art, Damien Hirst's pickled sheep, the photographic projects of Spencer Tunick and the mixed media 'installations' of artists such as Ai Weiwei and Cornelia Parker. As in other fields (see page 100) postmodernism in art is often playful, and seeks to undermine previously unspoken assumptions, broadening our understanding of what art is, and what it can say to us.

TALK LIKE A GENIUS

❛ Think of modern art, and you're probably just as likely to think of a video installation, a pickled shark or a pile of bricks laid out in a geometric pattern as you are of a painting. That all goes back to Duchamp, and whatever your reaction, the idea that anything can be art if the artist says it is vastly expanded the possibilities of modern art. ❜

❛ People like Warhol and Damien Hirst get criticized for the fact that they have assistants and a workshop-like approach to manufacturing art, but ironically you could say exactly the same about Titian and Rubens – there's a reason you see a lot of paintings attributed to "studio of ...". The idea of a solo artist sitting down and making a painting from start to finish was a bit of a romantic affectation, and Warhol's Factory was just a reversion to the usual practice. ❜

WERE YOU A GENIUS?

1 TRUE – when asked to name one of his paintings, Monet came up with the title *Impression, Sunrise,* which was later applied to the entire movement.

2 TRUE – Malevich's 'suprematism' flourished only briefly in his native Russia, but had a lasting influence.

3 TRUE – in particular they were inspired by the four-dimensional mathematics of spacetime.

4 FALSE – works by New York artists Lee Quinones and Fab 5 Freddy were exhibited in Rome as early as 1979.

5 FALSE – outsider artists include the poor, disempowered or mentally ill – people outside of the establishment, but not through choice.

Modern art starts out by trying to be something that photography cannot, before the conceptual revolution reveals that art can be almost anything.

History of literature

'It is only a novel ... or, in short, only some work in which the greatest powers of the mind are ... conveyed to the world in the best-chosen language.'

JANE AUSTEN

The origins of literature go back long before the written word, to stories told by our prehistoric ancestors around campfires. For most of history our written sources are limited – paper was expensive and books had to be handwritten by scribes, so effort was rarely wasted on preserving fiction. The invention of the printing press changed that, and the vast scope of great literature written since creates a daunting minefield even for the self-proclaimed genius.

If you can't actually find the time to be well read, then knowing the big movements and broad flow of literary history can at least help you appear that way.

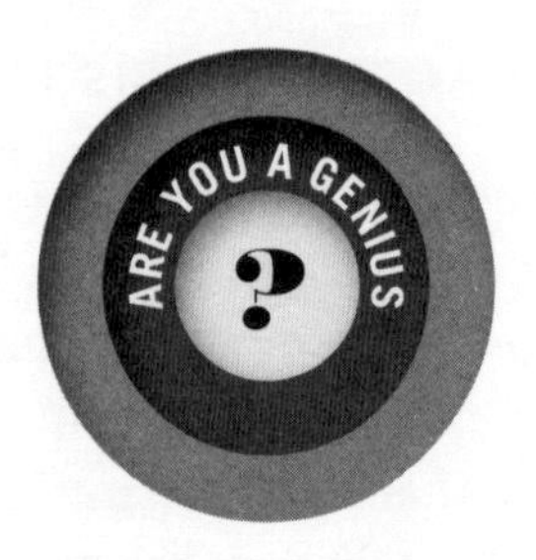

1 The *Iliad* is thought to have been made up by Homer long after the Trojan War it supposedly describes.

TRUE / FALSE

2 Literary historians have long been fascinated by a lost Shakespeare play called *Love's Labour's Won*.

TRUE / FALSE

3 *Sturm und Drang* was a German literary movement of the mid-19th century focusing on intense emotions.

TRUE / FALSE

4 James Joyce introduced the 'stream-of-consciousness' approach to writing in his modernist novel, *Ulysses*.

TRUE / FALSE

5 The *One Thousand and One Nights* is a collection of Arabian folk tales assembled some time around the 8th century CE.

TRUE / FALSE

TEN THINGS A GENIUS KNOWS

1 Epics and romances

From the *Epic of Gilgamesh* (*c.* 2100 BCE) through Homer's *Iliad* to *Beowulf*, the epic poem is the earliest form of literature. Such tales show traces of their origin in pre-literate storytelling, such as rhyme schemes to aid memory and enhance storytelling. Later poetic epics include Virgil's *Aeneid* (19 BCE, a politicized Roman foundation myth), countless medieval romances (chivalric epics, such as the tales of King Arthur) and complex allegories such as Dante's *Divine Comedy* (1320). Epics, romances and 'novellas' often incorporate shorter stories in a framing device; for example Boccaccio's *Decameron* (1354) and Chaucer's *Canterbury Tales* (*c.* 1400).

2 Plays and their influence

Scripted performance has its earliest origins in religious ritual, but the Western theatrical tradition is rooted in ancient Greece. The Greeks recognized three genres of play; tragedy (usually historically based), bawdy 'satyr plays', and comedy, which went through three major phases. 'Old Comedy' (exemplified by Aristophanes *c.* 400 BCE) was topical, political and viciously satirical. 'Middle Comedy' was more cautious and general in its targets, while the 'New Comedy' of Menander and others (*c.* 300 BCE) eschewed big politics for sitcom antics and the concerns of the ordinary man. Following the fall of Rome, theatre in Europe was once again dominated by religious ritual and instruction, but straight reenactments gradually diversified into mystery and morality plays, while a secular tradition reestablished itself through broad farces and plays based on folk tales.

3 Shakespeare

The influence of poet and playwright William Shakespeare on not only English but all later Western literature is unparalleled. His canon of 38 plays (including collaborations) frequently borrows stories and themes from earlier and contemporary works, but his unique sensibility adds astonishing depth, both in its investigation of the human condition and its subtle wit and wordplay. Both an actor and a businessman, Shakespeare thrived in the late 16th and early 17th centuries, becoming popular in the courts of both Queen Elizabeth and King James and often rewriting history to suit the political imperatives of the day. Spanning barely two decades, the exact chronology of his works is disputed, but there is a broad progression from early histories through crowd-pleasing comedies to major dramas (*King Lear*, *Macbeth*, *Hamlet*) and more complex 'late romances'.

4 Don Quixote

Spanish author Miguel de Cervantes changed the literary world forever with his *Don Quixote*. The first volume (1605) tells the comical story of Alonso Quixano, a deluded would-be knight who recruits hapless farmer Sancho Panza as his squire, in an epic work that lampoons traditional romances. Games and references to other works abound, with various chapters attributed to supposed archival sources in an early example of the technique known as 'metafiction'. The second volume (1615) is both more playful and psychologically deeper. The characters now encounter people who've read of their exploits, while Quixano struggles to recover his sanity while succumbing to illness. Some critics see *Don Quixote* as the first real novel; it's certainly one of the most influential.

5 Birth of the novel

Many cultures, including ancient Greece and Rome, Byzantium, China and Japan produced long-form prose works. However, the modern novel probably had its genesis in late Renaissance Europe (particularly Spain and Portugal), before rising to dominance in the 18th century. Novels are generally assumed to have a degree of psychological depth in their handling of characters, and to be written for personal reading rather than performance, but none of the rules are fixed, and the novel is an incredibly varied form. Significant early landmarks include Richardson's epistolary *Pamela* (1740), Walpole's *The Castle of Otranto* (1768, the first gothic novel, forerunner of the horror genre), Goethe's *The Sorrows of Young Werther* (1774), Jane Austen's *Pride and Prejudice* (1813), Mary Shelley's *Frankenstein* (1818) and Walter Scott's *Waverley* (1814).

6 From romance to realism

The 19th century saw an explosion in literature, driven by cheaper printing, wider literacy and appealing new approaches. In Britain, Walter Scott pioneered the historical romance, but it's Charles Dickens who truly dominates, with popular short stories and novels that deftly mix humour with darker themes and outrage against social injustice. Serialized publication reduced costs and increased readership, while cliffhangers turned books like *The Old Curiosity Shop* (1840–41) into the audience-teasing soap operas of their day. George Eliot equals Dickens in skill and complexity if not output, and her *Middlemarch* (1871–72) is a masterpiece of Victorian realism, while the Brontë sisters' handful of books from the late 1840s toy with concepts such as unreliable narrators and varying points of view. Dickens, in particular, was hugely influential in America, where domestic novels developed from the romances of James Fenimore Cooper (*The Last of the Mohicans*, 1826), through ambitious, richly layered works such as Herman Melville's monumental *Moby-Dick* (1851, notable both for its vast scope and its author's huge range of stylistic tricks), to the romantic realism of Henry James, whose interior monologue and psychological depth foreshadow modernism.

7 European novels

Though it's hard to generalize, European novels of the 19th century were perhaps more focused on the inner lives of characters than on labyrinthine plots. Pushkin's *Eugene Onegin* (1833) is an expansive portrait of Russian society, but also a description of a man's search for purpose that sets the stage for later Russian novelists, such as Dostoyevsky and Tolstoy. Flaubert's *Madame Bovary* (1856), a key work in French realism, meticulously dissects the psychology of a woman drawn to her doom by the stifling conventions of provincial life. Other French writers, such as Victor Hugo and Emile Zola, took up the British tradition of social agitation in works such as *Les Misérables* (1862) and *Germinal* (1885). Late in the century, Polish-born Joseph Conrad united European and British traditions in remarkable novels, such as *Heart of Darkness* (1899).

8 Modernism

Rooted in the early 20th century, modernist literature developed as a response to developments such as Freudian psychoanalysis, Nietzschean philosophy, and later scepticism in the aftermath of World War I. Poets such as Ezra Pound and T.S. Eliot produced densely allegorical, complex works with hidden layers of meaning, while novelists such as Virginia Woolf constructed novels with fractured, time-shifting narratives. Others pushed both literary and social boundaries; James Joyce's *Ulysses* (1920) reframes the *Odyssey* in a day's journey across Dublin that pioneers stream-of-consciousness, while D.H. Lawrence's *Lady Chatterley's Lover* (1928) broke both sexual and class taboos.

9 Genre

While the concept of genre (broadly denoting subject matter) as a means of categorizing literary works goes all the way back to ancient Greece, the modern idea of 'genre fiction' originated as a convenient marketing category for publishers during the paperback boom of the mid-20th century. In this sense, genre is fiction with settings or themes that are not strictly realistic, written to appeal to readers of similar works. Literary fiction, in contrast, was widely held to be grounded closer to reality, and focused on character rather than plot. Westerns, science fiction, fantasy, horror, thrillers and romance all get caught up in a catch-all category that is all too often used by snobs to distinguish it from 'proper' literary fiction.

10 Postmodern literature

In recent decades, the literary novel has shaken off some of its realist shackles as 'serious' novelists have been attracted to using the themes, tools and set dressings of genre for their own purposes. Science fiction has been a particular inspiration, and novels such as Thomas Pynchon's *Gravity's Rainbow*, David Mitchell's *The Cloud Atlas*, and Kazuo Ishiguro's *Never Let Me Go* all draw on its motifs. Other authors have flourished happily on the boundary between literary and genre (including Margaret Atwood, Doris Lessing and Iain Banks). Meanwhile, magical realism (Gabriel García Márquez, Isabel Allende, Salman Rushdie and others) in which apparently fantastical events occur in an ostensibly realistic world, bears many hallmarks of genre, yet retains literary kudos.

TALK LIKE A GENIUS

❛ The obsession with the dangers of "genre" goes way back – Cervantes has Don Quixote's niece burn all his chivalric romances because they're a bad influence on him, and Jane Austen hangs the entire plot of *Northanger Abbey* on a heroine who's read too much Gothic horror. That said, in both cases the authors are also laughing at the po-faced people who think the wrong kind of books are bad for you. ❜

❛ There's another school of thought about modernism that rather takes the shine off its inventors. Oxford critic John Carey reckons it was all about the intellectual class inventing a new form of writing that was inaccessible to the newly literate masses who hadn't had the "right" sort of education. There's certainly evidence that writers like Virginia Woolf and even D.H. Lawrence had a loathing for the lower classes that, at times, they didn't even bother to disguise. ❜

WERE YOU A GENIUS?

1 FALSE – the *Iliad* may have been written around 750 BCE, but it seems to preserve details from 400 years previously, passed down in oral form.

2 TRUE – although many suspect that it's just an alternative title for one of his other existing plays.

3 FALSE – the movement (which means storm and stress) lasted only from 1760–80. Johann von Goethe and Friedrich Schiller were both associated with it.

4 FALSE – Joyce brought it to a new level, but the technique, attempting to mimic our rambling natural thought processes, has precursors as far back as *Tristram Shandy* (1757).

5 TRUE – the collection also includes stories from Persian and Indian sources.

The Western literary tradition, founded on epics, romances, tragedy, satire and farce, reaches its full potential with the invention of the ever-changing, ill-defined entity we call 'the novel'.

Literary criticism

'Some books are to be read only in parts, others to be read, but not curiously, and some few to be read wholly, and with diligence and attention.'

FRANCIS BACON

While there's plenty of enjoyment to be gained in simply reading a book for its plot, characters and dialogue, literary criticism is the art of looking deeper. Critics investigate the stylistic tools the author uses to convey their meaning, and search for hidden or overlooked contexts (whether deliberately intended by the author, or inferred by the reader, it hardly matters). While it's possible to read and analyse nonfiction in this way, the criticism delivers its richest dividend when directed towards fiction – and certain kinds of fiction in particular.

Pay close attention – any would-be genius should be armed with some of the tools of critical theory.

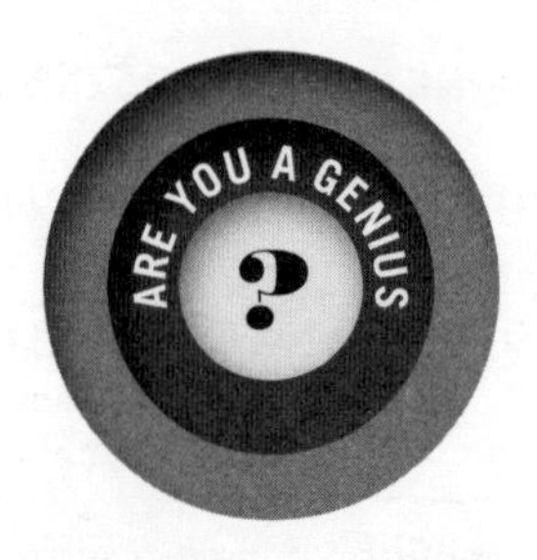

1 Arisotle's *Poetics* is the origin of the three-act structure found in many plays and movies.

TRUE / FALSE

2 Popular since the 1940s, so-called 'New Criticism' is a form of close reading of the text that deliberately ignores the author's own intentions.

TRUE / FALSE

3 Historicism, as used in criticism, is part of a wider philosophical outlook that influenced Hegel and Marx.

TRUE / FALSE

4 Feminist literary criticism emerged from the 'second-wave' feminist movement of the 1960s.

TRUE / FALSE

5 Literary Darwinists attempt to apply the ideas of evolutionary natural selection to the development of literature.

TRUE / FALSE

TEN THINGS A GENIUS KNOWS

1 Aristotle's *Poetics*

Literary criticism begins with the ancient Greeks, and in particular, Aristotle's musings on the nature of story. *The Poetics* (*c.* 335 BCE) focuses specifically on drama and poetry, and analyses the way in which poetic metre, musical melody, subject matter and presentation can affect our reaction to a work. At its heart is Aristotle's analysis of the tragic (dramatic) form; an argument that characters and plot should be illustrative of a central theme and the central aim of drama is to deliver 'catharsis' (a purification or release) to the audience. *The Poetics* have had a huge influence on subsequent criticism and literature itself – watch any Hollywood blockbuster and you'll see Aristotle's ideas at work.

2 Intent and reading

'Lit crit' typically involves looking beneath a story's surface elements of plot, incident and character, to analyse deeper messages and meaning. As such, we need to look at every aspect of a work, from individual word choice to overall structure, and messages both explicit and implicit. Some might argue that the author should have the last word on what their work means, and that criticism on this level is just a silly intellectual game. Indeed, despite popular misconceptions, few critics would argue for completely ignoring the author's intent – more often, criticism is aimed at seeing the techniques, both conscious and unconscious, that the author uses to achieve certain responses from the reader.

3 Diegesis and mimesis

The Greeks identified a division between two fundamental types of storytelling that still stands today. 'Diegesis' is a story as recounted through a narrator, while 'mimesis' is a story depicted through action, without ostensible narration. Broadly speaking, then, diegesis involves framing prose while mimesis involves a 'pure' combination of action and speech (as in a play or movie). Lit crit sometimes borrows the same principle to analyse prose; hence diegetic elements are those parts of a text that actually manifest in the story-world – action and dialogue that would survive if events were lifted off the page and played out mimetically. 'Extradiegetic' elements, in contrast, are parts of the storytelling style – word choices, style and other devices that do not alter 'events' but nevertheless affect the reader's perception.

4 Narrators

A key element of any work of literature is the narrative voice used to recount it. Early novels often claimed to draw from other sources, effectively bolstering the authority of their narrative voice as a 'true' account. The tradition has continued to the present day, though these days it's treated more as a literary game. Epistolary novels use letters to tell the story as a series of episodes from one point of view or another, while first-person novels limit our understanding of the story-world to a single narrator (who may or may not be entirely reliable). The omniscient narrator who sees all with an apparently clear (though sometimes jaundiced) eye became popular in 19th century literature, but even here the mask of objectivity sometimes deliberately slips, and we find ourselves asking who is telling the story, and why they are presenting this particular version of events.

5 The affective and intentional fallacies

Historically, critics have highlighted two common errors that we make when trying to judge a work. The 'affective fallacy' is the idea that a text's importance is primarily to be judged by how it makes us *feel* – everything from whether it elicits sadness, laughter or excitement to whether we fundamentally like it or not. The 'intentional fallacy' is our tendency to grasp for authorial intent – to analyse a text in terms of what the *author* means, rather than finding our own meaning in the words on the page. Yet both of these fallacies have their place; an emotional response to a text is an important cue, and for most readers the very definition of a good read. And authorial intent may be set aside in favour of judging the text on its own merits, but we'd be foolish to forget that it exists.

6 Marxist and other specialist critiques

A variety of different critical methods might seem at first like hobby horses for the obsessed, but can in fact offer interesting new ways of looking at a text. Marxist critique looks at works in terms of both the explicit role of money, and of the broader power relations involved, while feminist and queer critiques look at the attitudes to gender and sexuality, and black criticism looks at the role of race. All these critiques, and others, consider both the diegetic and extradiegetic content of the work itself (the explicitly voiced actions and attitudes of characters and narrator as well as those hidden in the unspoken subtext) alongside the wider context in which works were both created and consumed.

7 Literary quality

Most critics would say that 'literary quality' is the thing that marks out material worthy of serious attention from the vast majority of more functional writing (including not only most nonfiction, but also large amounts of plot-driven 'genre' fiction). So how exactly can we identify this mysterious quality? Unfortunately, it's rather a case of 'you know it when you see it' – a precise definition is hard to pin down, but likely indicators include the use of plot and characters to shed light in different ways on a particular theme (as per Aristotle), signs of careful sentence structure and word choice (rhetoric aimed at persuading the reader to perceive things in a certain way), and the presence of subtext, hidden meaning, metaphor and allegory (scenarios not meant to be taken entirely literally) inviting us to draw broader lessons from the text.

8 Historicism

Another important question is how much attention we should pay to the different historical situations in which a text was written, and in which we read it. It's undeniable that we have a different experience reading, say, Dickens' *The Pickwick Papers* as a single volume today, to that which readers did when picking it up as a monthly serial in 1836. Not only is there a difference in format, but also in expectation (today we know this is the first major book by the one-and-only Charles Dickens). We lack contextual understanding of society at the time that would have been innate to *Pickwick*'s first readers, and find ourselves groping to understand cultural references, but conversely benefit from knowing where Dickens and society went next. And even if it's not necessarily *fair* to impose our own modern standards on any author or their characters, we can't help doing it, and it can provide valuable insights.

9 Critiquing visual media

In the 20th century, movies and television overtook print to become the dominant form in which we consume fiction. Both are hugely collaborative endeavours, in which the text (in the form of a screenplay) is just one element. Outside of the experimental arthouse, most screenplay writing still maintains links to the basic ideas outlined by Aristotle, but how are we to assess the media as a whole? Most critics agree that the director is primarily responsible for the way material appears on screen, and so-called 'auteur theory' takes this responsibility to its ultimate conclusion, arguing that the director can be critiqued as the guiding hand behind every aspect of a movie in just the same way as the author is held responsible for a book. However, the complexities of performance, the changing limitations of technology and many other external pressures raise questions about how complete the director's control is.

10 Mass media

One last important thing to understand about criticism is that its inquiring attitude extends beyond the world of fiction. The complex nature of our modern world makes us increasingly reliant on mass media, such as newspapers, TV channels and social-media feeds, yet none of these relays a truly objective picture of what is going on in the world. Even in the best of circumstances, the selection of news stories is inevitably edited by the logistics of where a story is happening, whether pictures/reportage are available and what else is competing for the same airtime/ink. At worst, stories are completely ignored, presentation is skewed and 'alternative facts' are invented in the service of the political or financial agendas of proprietors and advertisers. Academics Edward S. Herman and Noam Chomsky have written at length about the importance of taking a critical attitude to the message the media present us with in books such as *Manufacturing Consent* (1988).

TALK LIKE A GENIUS

‘ Casual discussions about books usually focus only on the diegetic elements, acting as if the story had really happened. Critical reading, on the other hand, often focuses on the extradiegetic elements – places in which the author “shows their hand”, if you like. But of course you don’t want to go too far in making that distinction; every element of the story ultimately exists because the author decided to write it that way. ’

‘ You can sometimes take these things too far; the infamous postmodern critic Jacques Derrida once spent 80 pages discussing the use of the word ‘yes’ in a single passage of *Ulysses*! ’

‘ Plenty of famous authors have honed their craft as critics of others – for example Edgar Allan Poe did a brilliant dissection of Dickens’ *Barnaby Rudge* that probably inspired his famous poem *The Raven*. And even if you don’t want to write yourself, the detective-story element of figuring out how a book was put together, spotting loose ends and speculating about the author’s original plans can be fascinating. ’

WERE YOU A GENIUS?

1 FALSE – Aristotle has nothing precise to say about structure. In the 1st century CE, however, Roman theorist Horace recommended a five-act form that was followed until the 19th century.

2 TRUE – though the idea of ignoring authorial intent was not central to the original idea.

3 TRUE – the historicist outlook, which analyses both events and literary works in terms of historical context, is often contrasted with the structuralist approach of rule-based interpretation.

4 FALSE – feminist critiques of literature, such as Virginia Woolf’s 1929 essay *A Room of One’s Own*, go much further back.

5 TRUE – since the 1990s, Darwinists have established a new approach to understanding literature in terms of selection pressures and the spread of ideas.

Criticism is the art of taking a text apart to see how it works, and trying to find out why it was written in that particular way.

Structuralism and semiotics

'The idea behind structuralism is that there are things we may not know but we can learn how they are related to each other.'

CLAUDE LÉVI-STRAUSS

Whenever we read or consume any form of communication, we are in essence interpreting a series of rules, signs and symbols that convey meaning. Understanding exactly how this system of interpretation works is the driving force behind movements in literary theory, linguistics, psychology and philosophy with some bewildering names. Nevertheless structuralism, semiology and hermeneutics are all united in asking the same fundamental question in different ways: how do we communicate and convey meaning?

If you want to get the message of a single word or a complex text, you still need to know the rules behind it.

1 Semiotics is a way of understanding human communication in terms of signs and their meanings.

TRUE / FALSE

2 Belgian artist René Magritte made a particularly famous comment on semiotics in the form of a painting.

TRUE / FALSE

3 Vladimir Propp identified 31 basic story types in his analysis of Russian folk tales.

TRUE / FALSE

4 Colourless green ideas sleep furiously.

TRUE / FALSE

5 Claude Lévi-Strauss explained many anthropological puzzles in terms of an innate human tendency to make distinctions between binary opposites.

TRUE / FALSE

TEN THINGS A GENIUS KNOWS

1 The hermeneutic tradition
The idea of interpreting communication as a series of symbols goes back to Aristotle, who coined the term *hermeneutics* (from the Greek for 'translation') around 360 BCE. Hermeneutics was chiefly concerned with the interpretation of ritual texts, and most religions therefore have their own hermeneutic traditions. The idea of applying it more critically to analyse the origin of texts arose in Renaissance Europe. A famous early example was Italian Lorenzo Valla's demonstration that the *Donation of Constantine*, a supposedly Roman Imperial decree used to reinforce the power of the Catholic Church, was an 8th-century forgery. Valla's analysis of anachronistic word choices and phrasing proved highly influential; similar techniques have been used since the Enlightenment to analyse the authorship of the Bible itself.

2 Peirce's semiotics
The word 'semiotic' (meaning roughly 'observant of signs') was once used only in relation to medical symptoms. It gained its modern sense in the 19th century through the writings of American philosopher Charles Sanders Peirce, who discussed it as a 'formal doctrine of signs', the system by which we observe signs of all sorts and learn to attach meanings to them. Peirce broke down his semiotics into a system of three elements: the object that creates the sign; the sign itself; and the means by which a meaning arises from the sign (what Peirce called the 'interpretant'). Interpretants can themselves act as signs, requiring a further cascade of interpretation before a final understanding is reached.

3 Saussure and Morris
Swiss linguist Ferdinand de Saussure established his own influential branch of semiotics that he called 'semiology'. His dualistic approach focused more closely on the use of signs in society, and the divide between the spoken or written word (the signifier) and the mental interpretation it triggers (the signified). In his posthumously published *Course in General Linguistics* (1916), Saussure argued that signs can be arbitrary, with no necessary connection to the meaning they convey. Equally influential was US philosopher Charles W. Morris, who in the 1930s argued that semiotics had three major concerns: semantics (the relation between signs and their meanings); syntactics (formal relations among signs regardless of their meanings); and pragmatics (factors affecting the way that we interpret signs).

4 Structuralism
Saussure's work on signs led him to an approach to language known as structuralism. His 'structural linguistics' analysed elements of communication in a variety of different ways, specifically how groups of signs can be connected into a 'paradigmatic' set (whose signs are substitutable, for instance the nouns or verbs in a sentence), and chained together by syntagmatic relations (rules of syntax). Saussure's method marked a significant departure from earlier historically based approaches to linguistics (which had generally paid more attention to the way languages develop over time). He was also careful to make a distinction between *langue*, language-as-concept with all its attendant rules, and the everyday speech or *parole* that we actually experience.

5 Philosophy's 'Linguistic Turn'
Meanwhile, philosophers were pursuing their own ideas about language. Mirroring attempts to put mathematics on a sound logical footing (see page 149), Gottlob Frege and Bertrand Russell both pursued the idea that language was built on a system of references, connections between propositions and concrete facts that could, it was hoped, be supported by rigorous logic. Russell argued for a concept of 'logical atomism', suggesting that since our philosophical understanding of the world is built on language, that understanding can be broken down to the level of indivisible, logically provable facts. Russell's former pupil Ludwig Wittgenstein outlined broadly this sort of system in his 1921 *Tractatus Logico-Philosophicus*, often seen as marking a significant 'Linguistic Turn' in 20th-century philosophy.

6 Ordinary language philosophy
While his *Tractatus* of 1921 superficially appeared to follow Russell's idea of rigorous language

as a root to knowledge, Wittgenstein's later work developed a rather different approach. He came to a belief that many of philosophy's central puzzles arise not from reality, but from misapplications of language (often created by philosophical jargon itself). Chiming with many of the ideas of structuralism, Wittgenstein's later approach inspired an influential movement that became known as 'ordinary language' philosophy. He also pursued the idea of 'language games', particular sets or structures of rules and references we follow in specific contexts and which are not, therefore, universally applicable. In his posthumous *Philosophical Investigations* (1953), Wittgenstein argued that 'the meaning of a word is its use', a position that put him at odds with previous views that the meaning of a word lies (at least to some extent) in the real world object that it signifies.

7 Literary theory and structuralism

The apparent power of structural linguistics inspired widespread interest in the way that frameworks of rules, signs and symbols shape meaning. Wittgenstein's 'language games' were one consequence in philosophy, but it was in literary theory that structuralism really caught on. Critics set to work looking for patterns in works of literature, and highlighting the apparent semiotic meanings that accompanied every word choice. A school of 'narratology' arose that sought to identify and break down universal story themes and structures, beginning with Russian folklorist Vladimir Propp's 1928 *Morphology of the Folktale*, and arguably ending in countless popular books claiming to hold the secret formula for writing a modern bestseller.

8 Lévi-Strauss and structural anthropology

Another field where structuralism wielded considerable influence was that of anthropology (the study of humans within their various different cultures). Frenchman Claude Lévi-Strauss blazed the trail with his influential 1948 account of the rules surrounding kinship in the native Brazilian tribes he had spent time with during the 1930s. In doing so, he turned many previous assumptions on their heads by showing how kinship groups arose and remained strong through observance of structural rules similar to the unspoken ones within our own society. Lévi-Strauss's underlying argument that all cultures have underlying similarities remains hugely influential, even if some of his specific observations about the rules involved have since been undermined.

9 Roland Barthes and *S/Z*

Structuralist criticism reached its apogee in 1970 with *S/Z*, Roland Barthes's exhaustive analysis of *Sarrasine*, an 1830 short story by Honoré de Balzac. Barthes argues that the work's meaning emerges through the interaction of five distinct 'codes'. The hermeneutic code refers to secrets withheld and revealed in the text; the proairetic code refers to the way the narrative is constructed to create tensions and expectations; and the semantic code considers the way that word choice conveys meaning. The symbolic code, meanwhile, deals with overall subtext; the way the work handles its often unspoken themes, and the cultural code refers to the work's interaction with knowledge about the real world of the time. Together, these five codes form a powerful set of tools for analysing any text.

10 Criticisms of structuralism

Recent decades have seen the structuralist approach somewhat in retreat (or at least, diminished to just one of several techniques for interpreting meaning). US linguist Noam Chomsky's idea of universal grammar (see page 29) is sometimes seen as undermining Saussure's approach, though the truth is somewhat more complex; it was the idea that Saussurian structure is a product of behavioural learning (as promoted by US linguist Leonard Bloomfield) that Chomsky's concept of innate grammatical rules really displaced. Another criticism of structuralist semiotics is that they are ahistorical, suggesting that all meaning can be extracted from an analysis of rules and patterns in the present moment without any knowledge of how the system developed. The same can be said for the strictly structural approaches to anthropology and literary criticism; they may work as a system for analysing why certain actions and signs have meaning in the specific context of, say, a game of football, but a work like Stendahl's *Scarlet and Black* loses much of its meaning when shorn of historical context.

TALK LIKE A GENIUS

❛ If you want a good example of someone playing games with structuralism and semiotics, take a look at Umberto Eco's monastic whodunnit *The Name of the Rose*. Eco was a theorist as well as a novelist, and he deliberately created the book as an 'open text' with huge scope for interpretations. The title itself references about half a dozen different possible meanings while the monk 'detective' at the centre of the story is William of Baskerville – referencing both the Sherlock Holmes stories and philosopher monk William of Ockham. The fortified tower that holds the abbey library is constructed like a medieval map of the world, and there's a blind librarian called Jorge of Burgos in tribute to Eco's inspiration, the Argentinian writer Borges. ❜

❛ Wittgenstein famously wrote that "if a lion could speak, we could not understand him." In other words, even if lions had a language and we could translate its individual signs into English, we still wouldn't be able to grasp their meaning because we have no knowledge of the conceptual scheme that underlies them; any concept of lion-ness we do have is still a product of our own human point of view. ❜

WERE YOU A GENIUS?

1 TRUE – it offers insights into the way our choice of signs affects our perception of the world.

2 TRUE – Magritte's *The Treachery of Images* (a picture of a pipe with the words 'This is not a pipe' written beneath it) can be interpreted as a semiotic game.

3 TRUE – he also divided all the characters into just seven basic roles.

4 FALSE – Chomsky came up with this nonsensical yet syntactically correct phrase, to show the limits of some structuralist ideas.

5 TRUE – he found evidence in such features as the division of tribal villages into two opposing kinship groups.

Semiotics considers language and structuralism analyses meaning, but both argue that the methods of communication shape the reality we perceive.

Postmodernism

'The cultivated person's first duty is to be always prepared to rewrite the encyclopaedia.'

UMBERTO ECO

If the first half of the 20th century was defined by modernism in art, literature and philosophy, the second half was marked by a distinct shift known as 'postmodernism'. In contrast to the broad modernist trust in science, technology and progress, the postmodern attitude is one of scepticism – an all-pervading doubt. Postmodernism mistrusts sweeping generalizations, grand narratives and moral and historical absolutes in favour of relativism, constant reassessment and questioning. The movement has shaped art, society and political discourse in the electronic age.

Ask what postmodernism means, and you shouldn't be surprised to get multiple and sometimes contradictory answers – that's sort of the point.

1 The term postmodernism was being used as early as the 1880s, but only caught on in the 1950s.

TRUE / FALSE

2 Some Marxist critics see the postmodern outlook as the logical evolution of an advanced capitalist society.

TRUE / FALSE

3 US artist Jasper Johns' 1954 painting *Flag* is often regarded as the first postmodern artwork.

TRUE / FALSE

4 Postmodernism erupted into the mainstream following events in France in 1968.

TRUE / FALSE

5 The science-fiction form known as cyberpunk is often seen as a uniquely postmodern genre.

TRUE / FALSE

TEN THINGS A GENIUS KNOWS

1 The origins of postmodernism

Although many people pin the emergence of postmodernism as a cultural force to the late 1960s, its origins go back significantly further. Existentialist philosopher Martin Heidegger is often cited as a significant influence on later postmodern thinkers, specifically thanks to his concern (voiced in a 1953 essay) that modern society risked making science and technology the sole means of assessing our interactions with the world, lessening our value as human beings. Ultimately, Heidegger made a forceful argument for philosophy to remake itself through a rejection of rationalism, although these views did not become fully apparent until posthumous interviews were published in the 1970s.

2 Poststructuralism

The origins of postmodernism are linked to poststructuralism, a movement that emerged as a critique (though rarely an outright rejection) of mid-20th-century structuralist ideas. While structuralism claimed to show the true meaning behind an array of phenomena through analysis of their semiotic structures (codes of meaning), poststructuralists saw this as insufficient without historical and cultural background. Accepting that the way we interpret phenomena is always shaped by their wider context and our own assumptions, poststructuralists cast off faith in semiotics' ability to unlock a unique true meaning – a change of outlook that when coupled with Heidegger's rejection of rationalism as a route to truth, would pave the way for later postmodern philosophers.

3 Borges and the postmodern

Perhaps surprisingly, the roots of postmodern literature can be traced back well before these theoretical underpinnings. Ever since the invention of the novel, authors have toyed with unreliable narrators, shifting points of view and the difficulty of pinning down truth and meaning, but Argentinian Jorge Luis Borges took this to another level. His short stories, published in collections *Ficciones* (1944) and *El Aleph* (1949) blend literary games, questions of authorship, meaning and intent, multilayered structures, fantasy and existentialism. For example, *Pierre Menard, Author of the Quixote* (1939) is a purported literary review concerning a fictional French author who, after laborious research, recreates Cervantes' *Don Quixote* word for word, and is regarded as a genius for the new layers of meaning introduced by the mere act of rewriting the work in the 20th century. Borges is often seen as the progenitor of later Latin American 'magical realism' tradition but, as with his postmodernism, it's a label affixed with the benefit of hindsight.

4 *The Death of the Author*

Written on the cusp of the poststructural revolution, critic Roland Barthes's 1967 essay *The Death of the Author* cast a long shadow over postmodernism with its argument that a text can be interpreted without reference to the intentions of the author, and often in contradiction to their professed 'meaning'. The idea of 'close reading' (what some might consider a critical overanalysis of every individual word choice and sentence structure in a text) had become a standard of academic literary criticism since the 1940s, but much of this so-called 'New Criticism' was aimed at better understanding authorial intent. Barthes argued instead that once a text was surrendered to its readers, any interpretation gleaned from close reading of the text might be as valid as any other.

5 Michel Foucault

French philosopher and critic Michel Foucault played a key role in spreading postmodern ideas beyond the literary world and into the wider world of sociology, politics and historiography. Foucault developed what he called an archaeological method, explained in his *The Archaeology of Knowledge* (1969). He argued that the ideas and language used at any particular point in history (including the present) are inevitably shaped by the dominant subconscious assumptions and systems of knowledge of the time, and can only be properly understood in the light of those assumptions. Furthermore, he argued that discourses (stories of the development of ideas) presented by modern historians are simply narratives imposed on the past that break down under close inspection. Ideas held as 'fact' at any one time are therefore little more than consequences of a particular discourse, and can easily be replaced by others. In Foucault's view, then, the multiplicity of valid interpretations extends beyond the literary world of Barthes, and applies equally to any event, action or text however 'factual' it may appear to be.

6 Derrida and deconstruction

Philosopher Jacques Derrida rejected the maxims of structuralism as early as 1959, favouring the earlier tradition of phenomenology (the idea that we should approach cultural objects in terms of our experience and consciousness). He became famous for a critical approach that chimed with the ideas of Barthes and Foucault and became an important part of postmodernism. Derrida's 'Deconstruction' is a dismantling of preconceived notions about any subject in order to reconsider it with an entirely new approach. Its main technique is to identify the topic's inherent themes and then look for the elements within it that undermine them, revealing its fundamental conflicts.

7 Postmodern influences

Since its genesis, postmodernism has made its influence felt across a variety of fields, from art and literature to architecture and the social sciences. Constant reinvention, literal deconstruction, questioning of assumptions and a magpie-like attitude to previously strict categories (in everything from fashion to literature) are emblems of this wider postmodernism, and the word 'playful' is also frequently used, implying that no one should take any of this too seriously. In architecture, for example, postmodernism often involves structurally unnecessary adornment, detail and colour (the work of Frank Gehry offers many examples). Postmodern art, meanwhile, addresses questions about the limits of art with provocative answers, such as screen-printed soup cans (Andy Warhol) and monstrous balloon animals (Jeff Koons). Within the social sciences, postmodern psychology and sociology both focus on human behaviour and relationships in terms of communication and language frameworks rather than attempting to build 'grand narratives'.

8 Postmodern music

Music is another area where postmodernism wields a major influence. Classical modernism was defined by self-consciously 'difficult', complex and atonal works by composers such as Igor Stravinsky and Arnold Schoenberg. Postmodernism, in contrast, is associated with a minimalist style that emerged in the late 1960s through the work of Steve Reich, Philip Glass and others, characterized by repetitive simple melodies, rhythm and harmony. Pop music can also be seen as postmodern, both in its rapid turnover of competing genres, and its rejection of 'authenticity' (for instance, in the shifting personas of art-pop musicians such as David Bowie and Madonna).

9 Critics of postmodernism

As a movement, postmodernism certainly has its detractors; many serious thinkers regard it as at best useless in its inability to seriously address questions of knowledge, and at worst dangerous in its insistence that all attempts at definitive answers are equally valid and invalid. Noam Chomsky has raised questions around the inability of postmodernism to function as a system of analysis, while mathematician and physicist Alan Sokal became a *cause célèbre* in the late 1990s when he successfully submitted a deliberately nonsensical article to a supposedly rigorous cultural studies journal. Politically, postmodernism has been assailed by conservatives who see it as fostering moral relativism, and leftists who argue that its bourgeois games do nothing to address the real problems of society.

10 Post-postmodernism

Cultural commentators have been calling time on postmodernism since the 1990s, but what comes next? In architecture, some are detecting a return to more orderly approaches, while in literature, some critics talk of a hoped-for new engagement with sincerity. Arguments that 'something new' is happening are perhaps strongest in cultural studies, where the impact of the internet and social media is undeniable and dramatic. Here, some critics are concerned by an approach that engages with topics on only superficial levels, while others are more hopeful that our increased connectivity may lead to a deeper engagement with the world.

TALK LIKE A GENIUS

‘ The only thing that’s really new about postmodernist literature is the name. The idea itself has been around since Cervantes, but if you really want to see it in full flow, read Laurence Sterne’s 1759 *The Life and Opinions of Tristram Shandy*. Not only is it packed with references and allusions to contemporary politics and philosophy, it just outright plagiarizes long passages of other texts for its own use. It switches styles from one page to another, breaks the fourth wall to insult the reader, and even uses typographic tricks like different typefaces and layouts, and whole pages left blank or printed in black. As you’d imagine, then, it’s up there with *Ulysses* when it comes to tough reads. ’

‘ Scientists get edgy around postmodernism because its insistence that all attempts at knowledge are equally valid threatens the claims of science as a unique method of finding out the truth. You could certainly argue that the recent trend towards dismissal of experts – and equivocating about scientific fact in favour of controversy in areas such as climate change and vaccination – is a pretty self-destructive trend for a complex society that’s ultimately reliant on science and technology for its survival. ’

WERE YOU A GENIUS?

1 TRUE – the term postmodern was first used by an art critic describing the need to move beyond impressionism.

2 TRUE – US critic Fredric Jameson has argued that postmodernism marks an abandonment of the idea of progression, but that this is itself a foreseeable stage in capitalism.

3 TRUE – Johns’ encaustic painting of the US flag was widely seen as an invitation to multiple interpretations, paving the way for postmodern and conceptual art.

4 TRUE – many postmodern thinkers first rose to prominence after widespread civil and social unrest in May 1968.

5 TRUE – cyberpunk’s combination of cynicism, technological speculation and a magpie-like attitude to other genres is typically postmodern.

Postmodernism argues that there are many ways to interpret the world, all equally valid – a refreshing attitude that also carries hidden dangers.

Modern architecture

'I don't think that architecture is only about shelter, is only about a very simple enclosure. It should be able to excite you, to calm you, to make you think.'

ZAHA HADID

Environment is important to us and, as a result, architects wield extraordinary power over our lives; we literally can't get away from their work. What's more, architects seem to like nothing better than laying claim to a bold new approach or a break with previous conformity, which has led to a bewildering profusion of different movements and styles. It's little wonder, then, that modern architecture inspires such strong opinions, and if you want people to listen to yours, you'll need to know what you're talking about.

Should architecture be all about function, or does form matter too – and where do you draw the line between them?

1 The modern architecture revolution would not have been possible without mass production techniques.

TRUE / FALSE

2 The white exterior of New York's Guggenheim Museum was chosen to make it stand out when seen across Central Park.

TRUE / FALSE

3 A curtain wall is a structure that encloses a space but does not have to support any weight other than its own.

TRUE / FALSE

4 Sydney Opera House took 15 years to finish and was one of the first buildings to use computer-aided design.

TRUE / FALSE

5 Dynamic architecture is an increasingly popular design philosophy involving strong curves and organic shapes .

TRUE / FALSE

TEN THINGS A GENIUS KNOWS

1 Traditional architecture
Until the late 19th century, traditional materials placed physical limits on the nature of building design; wood and brick are relatively low in strength and so most domestic buildings remained small and functional, while stone, used on the largest buildings including churches, had limits to its strength, was relatively limited in distribution and cumbersome to transport. Following principles of 'statics' that had not changed since Roman times, architects pushed the laws of physics to build soaring church towers and vast domes supported with vaults, arches and buttresses that distributed forces and ultimately rooted them to the ground.

2 Technological breakthroughs
Recognizably modern architecture can be traced back to three 19th-century breakthroughs. First of these was the use of sheet glass, perfected by Joseph Paxton in the vast prefabricated pavilion that housed London's 1851 Great Exhibition (nicknamed the Crystal Palace). Second came the invention of the safety elevator by Elisha Graves Otis in 1851 – which for the first time allowed people to reach the upper storeys of tall buildings with ease. Finally, a breakthrough in the quality of manufactured steel in the mid-to-late 19th century opened the way for its use as a structural element. Stronger, lighter and more versatile than traditional masonry, steel frames were the key to a transformation in building techniques.

3 The Chicago School
Following a disastrous fire that destroyed much of Chicago in 1871, the rebuilding of the city became a testbed for innovative new architecture. In 1885, William Le Baron Jenney completed construction of the Home Insurance Building, a 'skyscraper' an unheard-of ten storeys tall, built around a fireproof steel frame that freed its masonry and glass walls of structural load. Other architects soon joined the frenzy for skyscrapers, and the work of the loosely bound Chicago School rapidly spread across America. Most influential among the Chicago architects was Louis Sullivan, who in an 1896 essay on the new style insisted that 'form ever follows function'. In other words, buildings should (and now *could*) be designed for their intended function rather than being limited by physics or stylistic concerns.

4 Modernism and futurism
Modernist architecture of the early 20th century followed Sullivan's maxim to the letter: buildings were designed around frameworks of steel girders, onto which internal divisions and external walls could be suspended. With less need to carry weight, the masonry element of walls could be largely replaced by much larger windows. Although early modernist buildings still made strategic use of decorative stonework, designs grew increasingly minimalist. Meanwhile, modernism spread beyond America – by the 1910s, Austrian-Czech Adolf Loos was attacking the whole idea of decoration in art and architecture, while Italian Antonio Sant'Elia was launching a bold manifesto for an aggressively modern futurist architecture, full of long lines and suggestions of speed and motion.

5 Frank Lloyd Wright
Probably the most famous architect of the 20th century, Frank Lloyd Wright (1867–1959) was apprenticed to Louis Sullivan in Chicago early in his career, but drew his own lessons from the idea of form following function. Wright often sought to design buildings that were modern but nevertheless in harmony with nature, using the full range of available materials to achieve his aims. Broad horizontal spaces were one of Wright's main themes; a famous example is Fallingwater, built in Pennsylvania in 1935 and cantilevered across a natural waterfall. By the 1950s, Wright was advocating an organic style that reached its ultimate expression in the elegant internal and external curves of New York's Guggenheim Museum.

6 Gropius and the Bauhaus
German architect Walter Gropius (1883–1969) was a key figure in functionalism, the wider movement that followed Sullivan's 'form follows function' maxim. From 1908 he worked in the office of Peter Behrens, the influential designer whose landmark AEG Turbine Factory in Berlin was the first

to combine glass walls and steel colonnades. In 1919, Gropius became director of the arts and crafts schools of the newly founded Weimar Republic, which he transformed into the hugely influential college of art, design and architecture known as the Bauhaus. Many talented European artists, architects and designers were recruited to teach at the Bauhaus, and in 1925, Gropius himself designed the school's landmark new building in Dessau, with an unprecedented glass 'curtain wall' along one entire side. Fleeing Germany after 1933, Gropius ultimately settled in New England, where he continued to produce influential buildings, such as the modernist Alan I. W. Frank House in Pittsburgh.

7 Gaudí and Le Corbusier

Although the secrets of concrete manufacture (lost since Roman times) were rediscovered in the 18th century, it was not until the 1890s that architects began to use it widely as a structural material. Antoni Gaudí's astonishing Sagrada Família church in Barcelona (begun 1882) made use of the ability to pour and cast concrete in complex shapes, ideal for the curvaceous art nouveau aesthetic of the times. Celebrated Swiss architect Le Corbusier (1887–1965) demonstrated the versatility of steel-reinforced concrete in a range of buildings throughout his career, the most notable being the controversial but hugely influential *L'Esprit Nouveau* pavilion, a stark white box with glass windows exhibited at the 1925 Paris Expo complete with cubist interior décor.

8 Late modernism

Following World War II, modernism continued apace, and the style known as 'brutalism' evolved in a rather extreme response to what some perceived as the blandness of earlier buildings. In place of smooth finishes, the brutalists made use of 'rough-cast' concrete to produce textured surfaces that were left unfinished, alongside exposed cement, brick and steel, with structural elements deliberately exposed. Brutalism was particularly common in large-scale housing projects, where its typical combination of modular apartments and high-rise 'streets in the sky' became a precarious social experiment. Some architects, however, found more elegant ways of using steel and concrete – most famously at the new Brazilian city Brasilia, designed by Oscar Niemeyer and built with astonishing speed in the late 1950s.

9 High-tech architecture

Also known as 'structural expressionism', the high-tech school of architecture follows the brutalist principle of exposing structural elements, but with startlingly different results. While brutalism tends to be all about the building's 'skin', high-tech proudly displays its skeleton. The 1950s geodesic domes of Richard Buckminster Fuller are an influential early precursor, but it was cantilever construction, with floors anchored to a concrete central core and supported further out by steel beams, that made high-tech possible by permitting lightweight glass and metal walls. Emerging in the 1970s, early high-tech tended to follow the minimalist blocky shapes of functional modernism, but designs soon became more exuberant as they made full use of the new materials and construction techniques.

10 Postmodern architecture

As early as the late 1960s, some architects began to kick back against the assumptions of restraint and minimalism inherent in the modernist approach. More ornate design features and the use of colour became popular, but it was the new techniques of high-tech architecture that really let postmodernism off the leash. Architects such as Robert Venturi and Philip Johnson played with introducing traditional architectural motifs to modern buildings in various US cities, while Richard Rogers and Renzo Piano transplanted service ducts, escalators and lifts to the outside of buildings such as the Centre Georges Pompidou in Paris in order to maximize interior volume (so-called 'bowellism'). In recent years, architects such as Norman Foster and Zaha Hadid have pushed the possibilities of modern architecture in new directions with expressive curved shapes that toy with conceptions of what a building should look like, while the firm of Skidmore, Owings & Merrill have taken buildings, literally, to new heights with their 'bundled tube' approach to the construction of skyscrapers.

TALK LIKE A GENIUS

❛ Probably the most famous brutalist architect was Ernö Goldfinger (1902–87), of course – mainly because he had a Bond villain named after him. It's safe to say that Ian Fleming wasn't a fan of his style! ❜

❛ The Greeks used to divide their columns into three elements: the base, the shaft and the cornice. It's amazing how that concept of a building or element in three parts has persisted into modern architecture. Even now, it's pretty rare to see a building that's exactly the same all the way up. ❜

❛ Le Corbusier famously said "a house is a machine to live in." You can see his point of view, but also why plenty of critics later attacked the idea; architects seem to like nothing better than a good maxim, and they did rather take the "machine" approach to extremes with social housing schemes in the sixties and seventies. ❜

WERE YOU A GENIUS?

1 TRUE – for example, the Crystal Palace was built in five months with prefabricated glass and iron.

2 FALSE – the white concrete was left exposed as a cost saving – Wright had originally intended the exterior to be dressed in stone.

3 TRUE – freed from the need to support the upper stories or roof, walls can be made from materials such as sheet metal and glass.

4 TRUE – computers were used to find an economic and structurally strong design for the famous roof.

5 FALSE – still rare in practice, it's actually the design of buildings that reconfigure themselves and change shape.

Architecture is always a compromise between vision, function and physics; its modern development has been driven by new materials that allow both physics and vision to go further.

Democracy

'VOTE, n. The instrument and symbol of a freeman's power to make a fool of himself and a wreck of his country.'

AMBROSE BIERCE

Democracy, in all its various forms, is the most widespread and enduring system of government in the world, yet it still attracts criticism and debate. At its heart, democracy's strength in inviting much of the population to have a say in government is also its weakness, in that to function properly, it requires real and authentic engagement from voters. Many of its controversies, meanwhile, arise from questions as to how the influence of voters should be represented, and how other actors, ranging from pressure groups to media organizations, attempt to shape the process.

Where does democracy come from, and how can we best organize it to fairly represent the needs of the people?

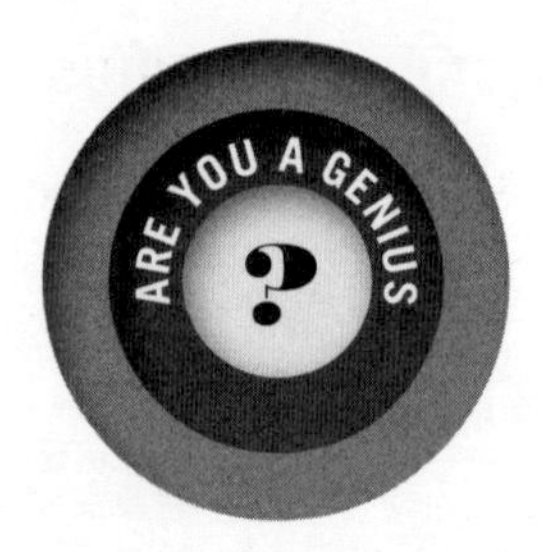

1 Jean-Jacques Rousseau argued that the social contract outlined by Hobbes had enabled the weak to tame the rich and powerful.

TRUE / FALSE

2 The US electoral college system was invented as a means of balancing power between populous and less-populated states.

TRUE / FALSE

3 Recall elections are a means for an electorate to force a new vote on whether their representative should continue in between national elections.

TRUE / FALSE

4 Proportional representation allocates seats in parliament in strict accordance with rankings given by the electorate.

TRUE / FALSE

5 There are no functioning direct democracies surviving in the modern world.

TRUE / FALSE

TEN THINGS A GENIUS KNOWS

1 Beginnings of democracy

Democracy started out as one of many competing systems of government practised among the city-states of ancient Greece – the word itself means 'power of the people'. Democracy is generally held to have begun in Athens after the people rose up to overthrow dynastic tyrants. Today, we would call that first system a 'direct democracy' – all citizens were de facto members of a legislative assembly with a right to vote on important issues, and citizens were chosen at random to fill various governmental posts. The most obvious failing of Athenian democracy to modern eyes is a limited definition of citizenship that excluded everyone except mature adult male landowners.

2 Representative democracy

First introduced under the Roman Republic, 'representative democracy' involves citizens voting for individuals who then (at least in theory) represent their interests in an assembly that actually votes on policies. This the most widespread form of democracy today, and has obvious advantages where larger populations are spread over wider geographical areas and direct democratic assemblies are impractical. Representative democracies generally allow people to get on with their daily lives with only occasional input into the democratic process, but as we shall see, they introduce a range of complications to an initially simple concept.

3 Forms of democracy

The majority of modern democratic states are republics with entirely elected leaderships. However, they still vary hugely in implementation. Some elect representatives to a law-making parliament or senate (which may have one or two chambers with different purposes). The largest political group then takes power to form an executive government that actually proposes legislation and runs the country. Others select a president in separate elections, and then expect them to appoint an executive (not necessarily elected) that will manage the various concerns of state and steer legislation through a parliament. A few countries remain constitutional monarchies, with royal families that have surrendered practical=yet still act as heads of state.

4 The social contract

But why do we need a government at all? Most modern thinking about this question is rooted in the ideas of philosophers Thomas Hobbes, John Locke and Jean-Jacques Rousseau. Hobbes' *Leviathan*, written at the height of the English Civil War, investigates the structure of society and comes down in favour of rule by an absolute sovereign power. However, it accepts that such rule cannot be imposed entirely by force; instead, Hobbes argued, individuals surrender to the sovereign because the alternative (a 'state of nature' characterized as a bloody struggle between individuals for basic survival) is too horrendous to contemplate. A generation later, Locke took a less dim view of humanity, arguing that the state derived its power by individuals voluntarily surrendering some innate rights in exchange for the protection of others fostered by the state itself (such as property rights and legal protections). In 1762, Rousseau coined the term 'social contract' for this fundamental deal that allows society to function; he saw the people as the sovereign element of the state, with government as an entity created by them for the necessary administration of certain tasks and roles.

5 Representatives

In most democracies, 'representatives' are expected to exercise a degree of judgement in votes rather than simply act as delegates to reflect the majority view of their electors. The theory is that the electorate gets a chance to pass judgement on the representative's performance at the next election and can get rid of them if they are unhappy. Unfortunately, this idealistic theory tends to be subverted by the business of party politics; independent politicians are few and most countries have political parties, groupings whose members and representatives share broadly similar principles, and which can deploy the necessary money and manpower to contest elections. Usually, therefore, representatives end up surrendering their independence to follow the party line, regardless of their constituents' wishes. What's more, the mechanics of the system often leave little opportunity to get rid of an unpopular representative if they represent a popular party.

6 Suffrage

Today, democracy and universal suffrage (the right to vote) feel like a natural fit, but this was not always so. Historical democracies often restricted the vote to males, landowners or those of certain religions, something that only really changed in the early 20th century. And even when everyone seems to have the right to vote, there are often barriers in practice, from the need for a permanent address or costly forms of identification, to the location of polling stations. Often these actions are legitimate attempts to reduce fraud, but sometimes they appear to disenfranchise certain groups. Countries such as Australia, in contrast, consider democratic participation so important that they make voting compulsory by law.

7 Electoral systems

Systems of election in different countries vary widely depending on geography, political priorities and historical accident. The United Kingdom and many US elections, for instance, use a 'first past the post' system in which the candidate wins if they command the most votes of anyone in a particular geographical constituency. This is said to ensure a strong link between constituents and their representative, but if there are more than two strong candidates, it can lead to the election of a representative without majority support. Many countries (such as France), therefore, implement a second-round 'runoff' between the leading candidates to ensure the winner has at least notional majority support. At the opposite extreme, pure proportional representation (PR) ensures a parliamentary balance exactly in keeping with the votes of the electorate, but sacrifices the constituency link and hands power to parties who control a list of 'approved' candidates. Compromises include regional PR (with larger constituencies electing multiple representatives, used in elections to the European Parliament), and single transferable vote (a system that redistributes second-preferences from less popular candidates until one achieves an outright majority).

8 Majoritarianism and rights

One significant problem in democracy is that the decision of a slim majority among the electorate (and sometimes not even that) can be used to pass laws that actively damage the opposition and their voters, or remove rights from certain groups. Such a system, known as 'majoritarianism', can lead to radical swings back and forth in policy, and as such most societies incorporate other branches of government, such as a strong and independent judiciary, a second revising chamber, or separation of powers between the executive and the legislature, in order to act as a safeguard against unrestrained executive power. Many countries have written constitutions to guarantee certain rights, which can only be altered in extraordinary circumstances or with consent of a legislative 'supermajority'.

9 Referenda

Sometimes a nation may find itself confronted by issues too large or controversial for the legislature, and in this case one solution is to resort to ask the electorate directly via a referendum or plebiscite. The popularity of referenda varies – Switzerland uses them as a tool of government without controversy, as do individual US states. Germany, in contrast, bans referenda under the constitution, viewing them as a route to dangerous populism. The need to boil complex issues down to simple yes/no questions is certainly a challenge, as is fair education of the electorate about the choices and consequences. As a result, many countries will only institute major changes based on referenda in the case of a significant majority – the alternative of a win by a narrow margin can leave a nation split down the middle.

10 Criticisms of democracy

While most democratic systems are great in theory, they often show problems in practice. The machinery of party politics is one issue, as is the question of funding; one person, one vote is all very well, but if political parties are reliant on private donations to support their campaigns, it's easy for wealthy individuals to buy influence. Lobby groups pressing for specific policies often employ ex-politicians, raising concerns about conflicts of interest, and the media – even when not biased by the views of its owners – plays a huge role in setting the parameters of 'acceptable' political debate. Perhaps the biggest challenge of all, however, is ensuring an educated electorate in the face of confusing messages on all sides and sometimes wilful ignorance.

TALK LIKE A GENIUS

❛ Freedom from rule by an absolute monarch or tyrant is pretty much the entire point of democracy, so it's little wonder the vast majority of democracies are republics. In some ways, that might be why the British monarchy survived – they gave up some of their powers to parliament from as early as Magna Carta in 1215. ❜

❛ Clearly everyone should have the right to register an abstention in some way or other, but one thing to be said for making voting compulsory is that it forces political parties to keep the whole electorate in mind, rather than just those segments that are most likely to vote. ❜

❛ A couple of political laws are always worth bearing in mind. Arrow's impossibility theorem is a neat logical proof that many of our concepts of a fair democracy are mutually contradictory; you can never come up with a voting system for choosing between three or more alternatives that is "fair" in every sense. Duverger's law shows how the voting system favours different party arrangements: first past the post tends to favour two-party dominance, while proportional representation allows many different parties to flourish. The question is really whether you want to put up with coalitions *between* parties, or coalitions *inside* parties. ❜

WERE YOU A GENIUS?

1 FALSE – Rousseau actually felt that the Hobbesian version of the contract favoured the powerful over the weak.

2 FALSE – the electoral college was originally a safeguard against Congressional scheming to install a President against the popular vote.

3 TRUE – however recall is not part of the electoral system in most countries.

4 FALSE – PR apportions representatives in line with popular support, but a ranked voting system is not necessary for this.

5 FALSE – two Swiss cantons retain the direct democracy model.

Winston Churchill once described democracy as the worst form of government except for all the rest, but that still leaves room to argue about which form of democracy is the least worst.

Conservatism, liberalism and socialism

'The two parties which divide the state, the party of Conservatism and that of Innovation ... have disputed the possession of the world ever since it was made.'

RALPH WALDO EMERSON

Three mainstream political traditions dominate democratic life in the Western world. Conservatism and liberalism have roots that go back to the late 17th century, while socialism is a comparative upstart with beginnings in the Industrial Revolution a century later. Despite their historical roots, however, the actual policies of groups that claim to espouse these beliefs are constantly changing. Sometimes they drive economic and social change, but at other times they merely shift to accommodate a new reality.

Natural hierarchy, freedom from government or democratic control and redistribution – which is best?

1 Conservatives believe in free markets and private enterprise, but also in traditional social values.

TRUE / FALSE

2 Socialists believe that in an ideal society, everyone would receive the same income.

TRUE / FALSE

3 Liberals believe in limiting the scope of government.

TRUE / FALSE

4 Communists believe in the common control of all property and a system in which everyone contributes and benefits according to their abilities and needs.

TRUE / FALSE

5 Anarchists believe in the complete elimination of the state, with government functions replaced by voluntary institutions.

TRUE / FALSE

TEN THINGS A GENIUS KNOWS

1 Whigs and Tories

The roots of modern Western political divisions can be traced back to England in the 1660s, when King Charles II returned to the throne in the aftermath of a Civil War. Two factions vied for power in Charles's parliament: the Whigs, who believed in limiting the monarch's powers, and the Tories, who argued that the monarch should be given more power. Nowadays, the Tory name lives on as a synonym for the British Conservative party, but the complex reality is that, while today's liberalism remains identifiably 'Whiggish' in its broadly progressive outlook, modern conservatism draws on both Whig and Tory thinking.

2 Modern conservatism

The modern form of conservatism is often traced back to the writings of Irish politician Edmund Burke (who actually sat in parliament as a Whig). Although a supporter of the American Revolution of the 1770s and a backer of early 'civil rights', such as lifting restrictions on Catholic landowners, Burke was horrified by the violent 1789 French Revolution. He later argued for the reinforcement of social hierarchy, religious and social traditions, and that along with rights come duties. Sardinian philosopher Joseph de Maistre put forth similar views around the same time that had a major influence in the Francophone world; the essential principle for both was that the ruling class is best suited to protect a country's legacy.

3 Locke and the origins of liberalism

The beginnings of the liberal tradition can be traced to philosopher John Locke's *Two Treatises of Government* (1689). Locke believed that government obtains its power from people who willingly surrender certain natural rights, with the state creating and protecting other rights in return. This view was hugely influential in shaping modern government (it forms the basis of the US Declaration of Independence), but the word 'liberalism' itself has since taken on other associations.

4 The rise of liberalism

In the 19th century, liberalism became a powerful force in politics. Coupling the economic theories of thinkers such as Adam Smith with Locke's earlier ideas about rights and philosopher Jeremy Bentham's utilitarian concept of ethics, liberals were broadly in favour of free trade and individual liberty. Radical democratic movements, such as early trades unions and suffrage campaigns, were taken into the fold, but there were tensions between competing interests. In 1859, liberal philosopher John Stuart Mill attempted to balance these conflicts with his book *On Liberty*, advocating a 'harm principle', namely that state power over individuals is justified only when it prevents those individuals from harming others. Mill also introduced the problem of the 'tyranny of the majority' (majoritarianism) and argued therefore for strong constitutional protection of individual rights. Mill's ideas had a huge influence on liberal politics around the world, although the precise boundaries of 'harm' have been subjects for much debate.

5 The origins of socialism

Arguments for a more egalitarian form of government have emerged from religions including Zoroastrianism, Islam and Christianity, and the Reformation and English Civil War both saw Protestant sects attempt to form socialist societies. However, the modern version of socialism, with its core belief that workers should own the means of production and a share in the profits of their work, is rooted in the Industrial Revolution. The rise of Napoleon stifled attempts to get egalitarian politics off the ground in France following its revolution, and so it was Britain that saw some of the first avowedly socialist groupings emerge and put their ideas into action. At first, socialist ideas were limited to the formation of so-called 'utopian communities' and 'workers co-operatives' – trades unions were banned by British law under a series of acts passed from 1799.

6 Politicized labour

Socialism gained strength as a political force in the mid-19th century. The appalling conditions suffered by both industrial workers and the agrarian poor led to the formation of various reformist movements, and a series of crop failures in the 1840s pushed things over the edge. Germany's Karl Marx and

Friedrich Engels published the *Communist Manifesto* in early 1848, and later that year (though largely by coincidence) revolutions erupted across Europe, resulting in the granting of significant new freedoms (not all of them lasting). Britain addressed discontent through a Great Reform Act (1832) that pacified the middle classes, but a mass movement called 'Chartism' continued to argue for political reform. Although the Chartists did not achieve their goals, the extension of suffrage to all men and most women in 1918 paved the way for the first Labour government a few years later. Today, such broadly 'social democratic' movements are the main left-wing political force in most countries.

7 Democrats and Republicans

The United States was founded on Lockean liberal principles, but its unique history has given rise to some peculiarities in its party system. Today's Democrats and Republicans both trace their origin to a party founded by Thomas Jefferson in 1791 to oppose Alexander Hamilton's Federalists. After overwhelming the Federalists, this party split in 1824 to create (among others) a socially conservative Democrat party and a more progressive Whig party, which absorbed many former Federalists and ultimately formed a broadly liberal Republican party. Today's complete reversal of these political positions is largely a result of three factors: former President Theodore Roosevelt abandoning the Republicans to pro-business factions in order to run as a third-party Progressive candidate in 1912; Democrat Franklin D. Roosevelt's astute 'New Deal' programme of the 1930s; and Lyndon B. Johnson's alignment of the Democrats with the Civil Rights movement in the 1960s.

8 Paternalistic conservatism

A long-lived and successful conservative approach to the challenges of both liberalism and socialism was 'paternalism' (sometimes known as 'one-nation conservatism'). While maintaining the class hierarchy, paternalists viewed their role in terms of a duty to provide for a stable society for the lower classes. Wealthy conservatives thus felt a need to cater for the needs of the masses through philanthropy (with a side order of moral guidance). This paternalistic contract, common in both Britain and America, allowed conservatives to appeal to a significant proportion of the working-class electorate and ensured frequent spells in government. In mainland Europe and Latin America, religious influence played a role in similarly motivated 'Christian Democrat' movements with broadly centre-right politics. For example, German Chancellor Otto von Bismarck established the world's first welfare state in the 1880s in order to garner working-class support that might otherwise have gone to socialist rivals, branding it 'practical Christianity'.

9 The post-war settlement

The aftermath of World War II left many European countries in a parlous state, with a widespread need for economic reconstruction and, in some cases, completely new systems of government. All nations saw the need for a peace dividend, not least in the face of a vastly expanded Soviet bloc. In Britain, Clement Attlee's Labour party took power in 1945 and instituted a welfare state and National Health Service that were seen as a model for other countries. The United States, meanwhile, went through an economic boom that ensured general satisfaction with economic conditions, while moving towards greater equality. This period of post-war consensus was closely linked to the Bretton Woods economic agreement (see page 133) and went largely unchallenged through four decades of government by various parties.

10 Neoliberalism and neoconservatism

The 'neoliberal' outlook that has dominated the politics of the past few decades, is at its heart an economic philosophy. Neoliberalism argues for a return to the core liberal principle of *laissez-faire* economics and minimal government intervention. Both President Reagan and British conservative prime minister, Margaret Thatcher, embraced these ideas and embarked on wholesale programmes of deregulation, privatization and government cuts, which continued under later governments that were ostensibly to their left, and spread around the world. Libertarians take these principles to an extreme, arguing for minimal government intervention in any aspect of business or personal life, while neoconservatives embrace the economics of neoliberalism but argue for curbs on 'progressive' social change.

TALK LIKE A GENIUS

‘ The idea of left and right wings comes from the French Revolution, when the National Assembly divided into supporters of the king sitting to the right of the president and supporters of the revolution to his left. The right didn't like the arrangement because they didn't really believe in political groupings, but they went along with it so they could hear themselves speak. ’

‘ The great success of neoliberalism was to convince everybody, at least for a while, that there was no alternative. Rival ideas about how to run an economy were pushed to the sidelines and mainstream economics reduced to a debate about the proper level of management, so political debate ended up being reframed around identity and social attitudes. You could argue these trends reached their inevitable conclusion with the financial crisis of 2008. ’

‘ Surprisingly, the first government to go heavily for neoliberal economics was the military junta that ran Chile under General Pinochet. The military didn't really know how to deal with the faltering economy when they took over, so they appointed a group of economists who'd all studied at the University of Chicago under Milton Friedman. By 1982, it clearly hadn't worked out well, but that didn't seem to discourage Thatcher or Reagan. ’

WERE YOU A GENIUS?

1 TRUE – although some streams of modern conservative thought have become more socially liberal.

2 FALSE – socialists argue for redistribution of income from the wealthiest to the poorest, but rarely to the point of eliminating differences in wealth.

3 TRUE – liberals argue for strict limits on government interference in individual liberty.

4 TRUE – but note that the Marxist communism attempted in the Soviet bloc is not the only possible form.

5 TRUE – however there's a stark division between traditional anarcho-communism and the anarcho-capitalist view that institutions can be replaced by market and contractual solutions.

The major political philosophies are defined by different attitudes to hierarchy, equality and liberty, but the perceived centre ground of debate is constantly shifting.

Digital politics

'Free speech is meaningless if the commercial cacophony has risen to the point where no one can hear you.'

NAOMI KLEIN

The rise of social media has the potential to transform politics, and has had a major influence on recent democratic votes around the world. Digital platforms promise to remove much of the power of old media by lowering the cost of publishing, turning the entire online world into a huge debating chamber. But at the same time those same platforms often put up new barriers that may shield us from uncomfortable truths. Any would-be genius needs a strategy for navigating this digital minefield.

The social media revolution shapes many different aspects of our lives, whether or not we engage with it directly.

1 Facebook has now overtaken broadcast media as the major source of news for people in the United States.

TRUE / FALSE

2 The word meme comes from the same ancient Greek root as imitation.

TRUE / FALSE

3 Publishing on the internet is only subject to the libel laws of the country where the material is hosted.

TRUE / FALSE

4 By the mid-2010s, US newspaper advertising revenue had returned to levels last seen in the 1950s.

TRUE / FALSE

5 Most countries have little regulation over online political advertising compared to printed forms.

TRUE / FALSE

TEN THINGS A GENIUS KNOWS

1 The medium and the message

Arguably, the first person to recognize the disruptive power of media technology was Canadian intellectual Marshall McLuhan. Writing as early as 1964, McLuhan described how the medium of transmission was just as important as the content it delivered, coining the memorable phrase 'the medium is the message.' McLuhan was concerned that our engagement with the overt content transmitted via a medium serves as a distraction from more insidious changes to our thinking and social interactions driven by the characteristics of the technology itself. In his terms, then, the significance of the social media revolution lies not in the individual messages it carries, but in the way it shapes society and our minds – democratizing publishing and allowing a platform to non-mainstream views on the one hand, but also shortening our attention spans and changing the sources of information that we rely upon.

2 Culture wars

One consequence of social media, however, has been an intensification of so-called 'culture wars', tensions between those with conservative and progressive outlooks that are almost as old as politics themselves, but which have intensified since the social liberalization of the 1960s. The phenomenon was highlighted in 1991 by US sociologist James Davison Hunter, who argued that differences in 'world view' across a whole range of social issues in America were becoming a more significant factor in politics than traditional divides of class, race and even political party. Up to this point, most politicians on all sides of these divides had broadly accepted the liberalization of previous norms, but Hunter pinpointed the emergence of a significant group that actually wanted to reverse these changes (while somewhat ironically also advocating an entirely modern form of deregulated economics).

3 Identity politics

The idea that your politics is defined by your self-identification as a member (by birth or by choice) of one group or another first arose in the US civil rights and anti-war movements of the 1960s. Most identity groups are on the political left, campaigning for social equality and against the status quo. However, such self-defining groups are sometimes criticized for being exclusionary and prone to splitting and factionalism (a phenomenon Sigmund Freud described as the 'narcissism of minor differences' as early as 1917). Identity politics is often attacked by conservatives who oppose social change, but some on the left also criticize it as missing the bigger picture of the struggle against inequality. Nationalism (see page 122) can also be considered a form of identity politics.

4 Memes

The term 'meme' was coined by evolutionary biologist Richard Dawkins as early as 1976, to reflect a parallel between Darwinian evolution and the way ideas spread in human society. Today, memes (often in the form of retweeted messages, captioned photos or short video clips) are one of the most common and fast-moving forms of social media communication. They propagate by their own form of natural selection; the most effective gain strength and spread widely through retweets and likes, perhaps mutating along the way by additional commentary, while weak ones falter as they struggle to find an audience. Just like a gene, a meme's strength indicates its suitability for survival and reproduction in a certain social environment. Of course, a meme's popularity does not necessarily indicate that it contains objectively true or useful information; it may just reinforce pre-existing prejudices, or simply be amusing.

5 Online monstering

One of the most disturbing aspects of online culture is the phenomenon of 'monstering', the rush to pour scorn on those who have overstepped a perceived boundary, either through real-world actions or online comments. This is nothing new; demonization has been a longstanding feature of traditional media, often in times of heightened tension. As a social phenomenon, punishment of those who break taboos probably originates deep in primate evolution (certainly, chimps do something very similar to punish rule-breaking individuals). The online incarnation is a form of meme, with outraged users spreading news of the original action or comment and inviting others to add their outrage. Like much of internet culture, perceived anonymity plays a role in making things more extreme.

6 Online and offline privacy

Interacting with any social media platform requires surrendering a degree of privacy. Platforms make money selling advertising, and target it using algorithms that build up a sophisticated profile based on a user's browsing history. Increasingly, however, these profiles are extending beyond the online sphere, with individuals tempted to surrender ever more information about their lives in the real world. Despite growing debate around the ability and power of government to collect data, it seems that many of us are willing to give up our most intimate personal information to private corporations in exchange for supposedly free services. As the saying goes, however, the reality is that if you're not paying for the product, you *are* the product.

7 The filter bubble

A unique aspect of social media is that it exaggerates our tendency to avoid dissenting views. Physiologically, having deeply held views challenged and undermined provokes very similar responses to being physically attacked. So most of us have an understandable desire to protect ourselves and avoid such confrontations, whether by subscribing to media that reinforce our world view or clicking the 'block' button. Search engines and social media, in their urge to feed us content that maintains our engagement (and, therefore, reaps advertising revenue), perform a similar task on our behalf without us even knowing it, such that many people exist in a 'filter bubble' that requires a conscious effort to escape.

8 The challenge to old media

The rise of social media and online news sources presents an existential challenge to traditional media, and the press in particular. The instant availability of reportage via social media can hugely outpace the sluggish realities of print, and the unlimited screenspace available means the digital realm can also offer in-depth analysis and reflection that was for a long time the unique advantage of newspapers over broadcast media. The withering influence of a handful of moneyed opinion formers may be seen as a positive result (though most newspapers have moved online, some new players with alternative viewpoints have established themselves for a fraction of the cost). But a reduction in funds available to pay for expensive but important investigative journalism (from any source) is certainly a negative one.

9 Targeted messaging

Since 2016, questions about the direct influence of Facebook, Twitter *et al* in democratic votes (most notably the 2016 UK Brexit referendum and US presidential election) have come to the fore. Some claim that social media profiling has now reached a point where information about a user's preferences can be used to identify their political allegiance and even the psychological 'pressure points' that might be used to make them change their vote. This allows individual voters to be targeted with tailored messages that, unlike mass-media advertising, are essentially invisible to outsiders and so are hard to counter or correct. The true potential of this approach is unclear, but the winning campaigns in those elections both spent large amounts of money with firms claiming to harness such technology.

10 Fake news

Fake news is properly defined as completely false stories that pander to the prejudices of a particular political grouping and thus spread as memes. Although often misapplied as a means of attacking real news stories that don't fit a preferred narrative, in its proper sense fake news is either planted by interested parties to derail political opposition, or made up purely in order to become memes and generate advertising revenue. Social media platforms are still getting to grips with the fact that they can be harnessed as political disruptors, and are striving to put measures in place for reporting fake news and countering its spread. There's also fierce argument as to whether they should bear the same responsibilities as traditional publishers, and just how far the right to free speech on the internet goes.

TALK LIKE A GENIUS

‘Is social media inherently biased towards the political left? There are certainly reasons to think it might be – for instance, research done in the UK showed that progressives were more likely to see it as a good thing (and, therefore, more likely to engage with it) while conservatives generally viewed it as a bad thing. If you start thinking that because your media feeds all agree on one thing, it's representative of the public as a whole, you can be setting yourself up for a fall.’

‘An early sign of the political power of social media was the Arab Spring movement that began in Tunisia in 2010 and rapidly spread across the Middle East. Apps helped news to spread rapidly outside of state-controlled channels, and were also used in the organization of protests – use of Facebook, for instance, soared. The story's more complex than that, however – Libya managed to have a revolution in 2011 despite extremely limited internet access.’

WERE YOU A GENIUS?

1 FALSE – although social media are a major form of news for the young, older demographics still favour traditional media.

2 TRUE – it comes from the word *mimeme*, meaning ‘a thing that is imitated’.

3 FALSE – although the law is complex and constantly changing, foreign hosting is not considered a reasonable defence.

4 TRUE – the fall in advertising since the millennium saw the closure of almost one third of US newspapers.

5 TRUE – although again, the law is constantly adapting to new concerns.

The digital revolution may have turned the world into a global village, but it's also created increased political polarization and even threatened to disrupt democracy itself.

Globalization and nationalism

'If globalization is to succeed, it must succeed for poor and rich alike. It must deliver rights no less than riches.'

KOFI ANNAN

Even before the widespread adoption of the internet, the postwar era had seen a vast leap in the world's interconnectedness or 'globalization', as economies around the world became ever more dependent upon each other and people travelled more freely. Nationalism, in contrast, was widely held to have been responsible for the devastation of World War II and was widely in retreat. Since the economic crash of 2008, however, the central assumptions of globalization have come under question as never before.

Globalization is an unavoidable fact of modern life, with both positive and negative consequences – the question is how we deal with it.

1 International agreement on standardized shipping containers is thought to have boosted trade between industrialized nations by around 700 per cent.

TRUE / FALSE

2 The 'triangular trade' was an early form of globalized commerce linking Spain, China and South America.

TRUE / FALSE

3 Although the United States is still the world's richest nation, its share of the global economy has fallen from just over one third to less than one sixth.

TRUE / FALSE

4 In the past 40 years, the number of people subsisting on less than a dollar a day has halved.

TRUE / FALSE

5 The Chinese economy is now 20 times larger than it was in 1979 when the country decided to pursue globalization.

TRUE / FALSE

TEN THINGS A GENIUS KNOWS

1 Origins of globalization
Globalization is nothing new; the Romans were doing it some 2,000 years ago, both through the exertion of military might to expand their physical empire, and through trading links that stretched to the edges of the known world. The Renaissance age of European exploration and conquest brought about a new wave of global trade and colonization, and the ensuing age of empires revealed a particularly ugly aspect, in the plundering of less developed countries for their resources. When we talk about globalization today, however, we tend to mean the global web of free trade between more and less developed nations that has arisen since the end of Word War II.

2 Modern globalization
In the immediate aftermath of World War II, most Western governments became signatories to the Bretton Woods System, which steadied the flow of investment capital between economies in order to maintain stable exchange rates and encourage investment into domestic economies. At the same time, however, many nations agreed to lowering trade tariffs in order to stimulate economic growth. When Bretton Woods broke down in the early 1970s, capital was able to flow more freely between economies, but lower tariffs remained in place. It therefore became much more attractive for multinational companies to manufacture goods in low-wage economies and import them to sell to wealthier customers at home.

3 Workshops of the world
A major step forward in global trade came with a surprisingly simple innovation – the standardized shipping container. Containerization brought with it huge benefits in terms of shipping costs, and vastly reduced demand for labour in the shipping industry, while encouraging overseas manufacturing. China's 1979 decision to pursue a market economy opened up international access to a vast new workforce, but also unleashed an entrepreneurial spirit that saw Chinese businesses begin to make use of their trading advantages themselves. Japan had already transformed itself from a workshop for Western manufacturers to a world leader in technology and innovation, and the same thing has since happened in South Korea and, to a lesser extent, India.

4 Multilateral organizations and agreements
Bretton Woods put in place two major pillars of the global financial system: the International Monetary Fund (IMF) and the International Bank for Reconstruction and Development (now the World Bank). Initially designed to regulate movement of capital, IMF members contribute to its funds and can be bailed out during financial difficulties, while the World Bank boosts investment in developing countries. The rise in monetarist economics saw both organizations encourage free-trade agreements made under GATT (the General Agreement on Tariffs and Trade) and its successor, the World Trade Organization (WTO). These not only lowered tariffs on physical goods, but also opened the way for deregulation of trade in intangibles like service industries, leading to the widespread phenomenon of outsourcing.

5 Globalist economics
Before the financial crisis of 2008, globalization saw an increase in trade and capital flows between countries (i.e. the purchase of goods, and the effective export of money in exchange). John Maynard Keynes argued that large trade imbalances (either the surpluses run by net exporters or the deficits accrued by net importers) are inherently destabilizing and both debtors and creditors have an obligation to reduce them, but his proposed scheme to do so was rejected in the Bretton Woods agreement of 1944. Nevertheless, a general tendency to view trade deficits as A Bad Thing lasted until the 1980s, when Milton Friedman put forward an argument that deficits are self-correcting in the long run. Released from what had previously been considered an important aspect of economic management, developed governments embarked on a deficit binge. For example the USA, on paper, now owes more than a trillion dollars to China.

6 Protectionism
Regardless of which economic theory you follow, one concrete downside of globalization is the tendency for developed economies to haemorrhage

jobs in sectors such as manufacturing, and for industries in less developed countries to outcompete those in advanced ones. Increased barriers to trade in the form of tariffs and quotas might seem an instinctive reaction when faced with such threats, but economists argue that protectionism is almost always the wrong response, with consequences worse than the problem it's trying to fix. Most obviously, it tends to raise the prices of both imported and domestic goods, leading to inflation (for example, the 'corn laws', abolished in 1846, kept British bread prices high in order to protect farmers). Just as problematically, protectionism can provoke a response in kind, leading to trade wars and slower economic growth for all concerned (something that played a key role in prolonging the Great Depression of the 1930s).

7 Immigration and work

Probably the most controversial aspect of free trade is the question of its extension to the labour market: should people be allowed to freely emigrate to find work in other countries? The straightforward economic answer is an unequivocal 'yes'; labour is just another type of economic good and so everyone benefits from its free exchange. Large-scale surveys have confirmed that immigrants help raise a country's overall GDP and do *not*, on the whole, depress employment or wage levels among natives (in contrast to the globalized trade in goods). However, immigration is not a purely economic issue; it can also increase demands on public services, and create cultural pressures in communities where it is concentrated. Tax raised from incoming workers should always be more than enough to pay for expanded services, but government still needs to ensure this increased income is spent in the right areas and, for reasons of ideology or sheer inefficiency, this is all too often not the case.

8 Multiculturalism and monoculturalism

The ways in which countries treat recently arrived immigrants varies hugely depending on both their deep history and their current political governance. Britain and the United States have historically favoured multiculturalism, allowing immigrants to retain ethnic and religious traditions from their countries of origin provided they live within the laws of their adopted home. France, conversely, maintains a monocultural approach born out of the French Revolution's principle of equality, with no special allowances for incomers, most notably in aggressively secular policies against Muslim dress. Both approaches have been accused (usually by supporters of the other) of creating alienated immigrant ghettoes and discouraging full participation in national life.

9 The crisis in globalization

Across Europe in particular, the past decade has seen a marked backlash against globalization, as the aftermath of the global financial crisis has left some inherent shortcomings cruelly exposed. Most significant among these is the problem that, while global trade does a great deal to equalize wealth between countries and improve the lot of the world's poor, it's the lower-paid workers in advanced nations who tend to lose out as their industries move abroad. This problem could be ignored while steady growth ensured that decent new jobs were being created to take the place of the old ones, but when the crisis hit in 2008 (its spread exacerbated by the deregulation of financial services around the world), the unspoken contract between neoliberal, pro-globalization governments and their working classes failed.

10 Anti-globalization movements

The heyday of globalization saw a huge increase in the interconnectedness of global economy and culture, with nationalism (at least in its most toxic forms) apparently retreating in favour of globalism. The 2008 crisis saw a reversal of this trend, with widespread protests against the consequences of globalization and a resurgence of populist political groups aiming to ride a wave of discontent with varying degrees of success (as seen, for instance, in the French Front National and Italy's Five Star Movement). Right-wing groups are concerned with issues of sovereignty and immigration (despite the fact that their leaders are often the very people who have reaped globalization's rewards).Left-wing groups remain broadly internationalist in outlook, but nevertheless argue in favour of workers' rights and better regulation (particularly regarding the ability of companies to ship jobs abroad and avoid taxation on profits).

TALK LIKE A GENIUS

‘The Silk Road wasn’t just one thing, but a whole network of roads and trading posts connecting east and west across Central Asia. Very few people travelled along it from one end to another, but you could usually find buyers who’d take your goods because they knew where to sell them somewhere further along.’

‘Multicultural societies get accused of creating ghettoes because people with similar backgrounds, beliefs and even diets prefer to live close to each other and near to their kinds of shops and restaurants. Some parts of Europe, on the other hand, have ended up with ghettoes through sheer economics; treating everyone equally means not giving special help to those who might need it in order to integrate, so they end up as an underclass languishing in housing projects.’

‘There’s a distinct tension in the fact that economically *laissez-faire* Western political parties who drove globalization when it was all about allowing manufacturing to go abroad, are often the ones who are most concerned when the consequences come closer to home in the form of imported immigrant labour.’

WERE YOU A GENIUS?

1 TRUE – this is the best estimate based on trade figures between the 1960s and 1990s.

2 FALSE – the triangular trade was actually the exchange of slaves, manufactured goods and tobacco between Africa, Europe and the Caribbean from the 18th century.

3 TRUE – although the US has continued to grow in real terms.

4 TRUE – although the actual living costs have risen steadily so more than a billion people remain in dire poverty.

5 FALSE – the Chinese economy is actually *40* times larger than it was before globalization.

The globalized world is a creation of technological changes reinforced by deliberate economic decisions. Attempts to reverse it could be disastrous, so we need to distribute its benefits more fairly.

Capitalism

'The new community which the capitalists are now constructing will be a very complete and absolute community; and one which will tolerate nothing really independent of itself.'

G.K. CHESTERTON

Capitalism is an approach to running economies based on the idea that individuals and privately owned businesses interact in markets. Despite rarely making conscious decisions about the larger-scale requirements of society, the theory is that these private actors will, nevertheless, ensure an efficient distribution of goods and money purely by acting in their own self-interest. Although capitalism is now almost universally adopted around the world, it comes in many flavours, often distinguished by differences in opinion on the role of government.

Capitalism underlies the workings of the modern world – you may not agree with it, but you really need to know how it works.

1 In economics, capital refers to the total amount of cash money circulating in the economy.

TRUE / FALSE

2 Under the fractional reserve banking system, banks are only required to hold enough cash to cover a small percentage of their savers' deposits.

TRUE / FALSE

3 The principle of comparative advantage states that a country should always strive to manufacture products that it can produce more efficiently than its neighbours.

TRUE / FALSE

4 A central bank's main roles are to hold the debt of its national government and issue bank notes.

TRUE / FALSE

5 Bullionism was an early financial model that measured a country's wealth purely in terms of its control of precious metals.

TRUE / FALSE

TEN THINGS A GENIUS KNOWS

1 The birth of money

Early money mostly consisted of coins minted in precious metals that carried inherent value, so-called 'commodity money'. However, governments soon learned the trick of debasing currency (reducing its weight or adding base metals while retaining the same face value). This had the natural effect of inflating prices, but could be useful for a government with a lot of debt to pay off and limited resources. Paper money, meanwhile, originated in China as early as the seventh century. Paper notes could theoretically be redeemed for a fixed amount of commodity money, but proto-economists soon realized you didn't necessarily need an equivalent amount of gold to back the stated value of all the notes issued, since few were likely ever to be presented for redemption.

2 The mercantile revolution

Modern capitalism is rooted in the mercantile system of late medieval England. Before this, money played a limited role in society; most countries tended towards a manorial system (peasants or serfs owed a bond of loyalty to the lord of the manor, who allowed them to farm the land in exchange for a proportion of their crops). That system came under fatal strain after the population crash of the Black Death (*c.* 1350) left landowners desperate for labour and gave peasants the upper hand. Many left the land for towns where they were free to sell their labour or engage in trade. Others stayed on, but became tenant farmers, selling produce for money and paying cash rent. The switch from subsistence farming to trade spurred innovations that improved productivity and allowed the economy to grow significantly.

3 Trading companies

The age of European exploration in the 16th and 17th centuries saw important innovations in trade emerge from the Netherlands and England. Probably the most significant of these was the 'publicly traded company', as exemplified by the Dutch East India Company, or VOC (founded 1602). Previous business enterprises too expensive for one person to finance had either taken the form of informal partnerships or 'joint stock' companies. Stock certificates entitling the holder to a share in the profits could be bought and sold in private, and in the event that the company failed, the shareholders' liability was limited only to the value of their stocks. While the VOC retained these benefits, it added the revolutionary idea that shares could be publicly traded on an open market. The birth of the stock exchange unleashed a much more powerful system for financing costly enterprises.

4 Central banking

The first state-owned central bank was founded in Sweden in 1668, but the model set by the Bank of England after 1694 proved more influential. Unable to borrow the staggering sum of £1.2 million to build a new navy from private individuals, the government came up with the idea of borrowing from the public en masse, and making lenders shareholders in a limited company. The new bank would hold government bonds (IOUs) as its primary asset, and be authorized to print bank notes secured against them. Over time, the bank's role in management of the British economy expanded significantly, and it became a model for the establishment of many similar central banks.

5 Market bubbles

Speculative bubbles are an all-too-familiar feature of capitalism that occur when people assume the price of an asset will go on rising forever, and therefore purchase purely with an eye to selling at a profit in the future. The asset's price rapidly loses any relation to its intrinsic value, rising rapidly as further demand is drawn in, before eventually reversing in a sudden collapse that can trap late entrants in a catastrophic loss. Famous bubbles include the Dutch tulip mania of the 1630s, the South Sea Bubble (affecting stocks in a proposed trading venture in the 1720s) and the 'Dotcom' bubble of the late 1990s.

6 *The Wealth of Nations*

Scottish philosopher Adam Smith is usually seen as the godfather of capitalism. His principal concern in *The Wealth of Nations* (1776) and other works was to argue that countries grew in wealth and power not simply through the acquisition of resources such as gold and silver, but through the efficient deployment of capital (the assets of individuals, including money, machinery and land, that can be used to generate profit). In particular, Smith argued for division of labour – the idea that if a manufacturing process requires three distinct steps, it's far more efficient to have one worker concentrating on each step than to have all three trying to take the process from start to finish.

7 Ricardo's insights

English economist David Ricardo established two key rules of classical economics in the early 19th century. He argued that the value of goods was directly related to the labour put into their manufacture, rather than their use to the purchaser (an idea that remained central until 1890, when Alfred Marshall showed the effects of supply and demand on price). He also discovered the principle of 'comparative advantage', the fact that if producers concentrate on manufacturing the goods they find most efficient and trade the resulting surpluses, everyone can be better off. This idea lies at the heart of arguments for free trade, and although it seems unlikely at first, the example below shows the principle in action.

Country:	Portugal	England
Labour hours per unit cloth:	90	100
Labour hours per unit wine:	80	120
Labour hours for 1 unit of each:	170	220
Total goods produced:	1 cloth 1 wine	1 cloth 1 wine
Concentrating effort in most efficient industry:		
Labour hours:	170	220
Total goods produced:	2.15 wine	2.2 cloth
If countries now exchange half their production through trade, each can end up with 1.075 units wine and 1.1 cloth for the same effort that previously produced 1 unit of each.		

8 Karl Marx

Although usually painted as the arch-enemy of capitalism, German political philosopher Karl Marx offers important insights into the way the system works. *Das Kapital* (1867–94) was built on Ricardo's labour theory of value, and argued that capitalists would always seek to maximize the surplus value (profit margin) on their goods, since this was money that they could reinvest to make further profit. Marx believed they would do this by paying workers ('the proletariat') as little as possible and using the threat of unemployment to keep them in line. He predicted this would ultimately lead to a proletarian revolution in which the workers overthrew the capitalists and seized the means of production for themselves, but failed to foresee that workers would be able to gain a stronger hand and improve their conditions through unionization and collective bargaining.

9 Fiat money

The vast majority of money in circulation today is called 'fiat' money (from Latin meaning roughly 'let it be so'). Instead of being backed by bullion (a practice that effectively ended when the dollar abandoned the 'gold standard' in 1971), the government and central bank permit it to be conjured up by a process called fractional reserve banking. The central bank injects money into the economy by lending to other banks or purchasing assets from them. Banks then lend this money out into the wider economy while keeping a small fraction in reserve. The lending does not increase the size of the economy in monetary terms (since it has to be paid back), but it does increase the amount of money in circulation. What's more, as the lent money is spent, it re-enters banks as customer deposits, and most of it can be lent all over again.

10 The modern financial system

Fractional reserve banking and fiat money are at the heart of modern finance, but have they got out of hand? The free flow of money they release allows financial institutions to create markets far larger than the underlying economy – for example, while globalization has boosted concrete foreign trade (the actual exchange of goods) to around $50 billion dollars per day, the foreign exchange markets (which ostensibly exist to service this trade) are 100 times larger. This is just one example, and the imbalance only gets more extreme as institutions develop and trade in ever more complex financial instruments. The crisis of 2008 was a salutary lesson in what can happen when such financial engineering goes wrong.

TALK LIKE A GENIUS

❛ Economists still dispute the cause of bubbles, and some even say they're just part of the natural economic cycle. The general view, though, is that they're a social phenomenon, and they may be more common when the assets in question represent a supposed paradigm shift in the way we do things – that was certainly the case with the Dotcom bubble and the South Sea Company. I guess this encourages people to think "it'll be different this time," but of course it never is! ❜

❛ People dismiss Marx at their peril – the main thing he got wrong was the idea that it was all going to end in tears, because he wouldn't countenance the idea that capitalists and governments might work together to raise the standards of living for the proletariat. ❜

❛ Adam Smith also introduced the idea of the "invisible hand", the self-organizing efficiency of markets. This concept goes largely unquestioned by modern economists, but it's startling to realize how much market incentives organize the flow of goods without any central control; it's why if you go to the shops for a pint of milk you'll be able to get one, but the shop rarely has to ditch milk because its date-life has expired. ❜

WERE YOU A GENIUS?

1 FALSE – for most economists, capital is actually money or goods that are invested in order to generate revenue in some form or other.

2 TRUE – banks rarely hold a large amount of their assets in cash.

3 FALSE – comparative advantage actually says that the country should strive to manufacture products that it can produce more efficiently than others within its own economy.

4 TRUE – although modern central banks play many other roles in monetary policy.

5 TRUE – bullionism was widely assumed to be correct in Europe during the Renaissance.

Capitalism is all about having the means to produce goods and make profit – but our ideas about how to do that have changed over the centuries.

Macro- and micro-economics

'It is not from the benevolence of the butcher, the brewer, or the baker that we expect our dinner, but from ... their own interest.'

ADAM SMITH

The overall economic situation of a country, firm or individual arises from a bewildering mix of forces acting on many different levels. In 1933, Norwegian economist Ragnar Frisch made a useful distinction between macroeconomics (the international, national and regional factors that most influence government decisions) and microeconomics (the small-scale factors that most influence individuals and businesses). Economists differ widely on what some of these factors mean and how they should be controlled.

Economics is a jungle of different concepts – do you know your exchange rates from your efficient markets?

1 In classical economics, the price of a product is assumed to fall when demand rises.

TRUE / FALSE

2 Pareto efficiency is a state in which goods are distributed between parties depending on how much benefit they will get from them.

TRUE / FALSE

3 Economics is built on the belief that everyone participating in a market does so while in possession of complete information, and always acts rationally.

TRUE / FALSE

4 Economists generally believe there needs to be a certain level of unemployment in order to prevent inflation.

TRUE / FALSE

5 When more money enters an economy, prices tend to rise in order to soak it up.

TRUE / FALSE

TEN THINGS A GENIUS KNOWS

1 Microeconomic definitions

Economists tend to use everyday terms in highly specific ways, so it's worth noting a few of them here. First, the 'value' of an economic good (a product or service that is brought or sold) is an indication of its inherent worth; the cost of physically creating it including, not just raw materials, but also labour and even intangibles such as intellectual value. A good's 'cost' is the amount of money expended in its manufacture, while its 'price' is the amount of money it can be sold for (hopefully equal to the cost plus a profit margin). Microeconomics is concerned with balancing prices to reflect market conditions.

2 Price, supply and demand

The price of goods follows some fairly simple economic laws. The law of demand says that demand for a product rises when its price falls, and falls when the price rises. However, this is not foolproof; goods that are necessary and cannot easily be replaced by alternatives tend to have 'inelastic demand', selling in the same quantities in all but the most extreme price changes. A related rule links supply (the amount of a good available to be sold) with demand. At any given price, supply and demand should find equilibrium, so just enough goods are provided for those who want to buy them, and every seller can find a purchaser. When supply falls, prices will rise, lowering demand but also stimulating production so that equilibrium is restored. Conversely when supply rises, prices fall due to a lack of demand, and production is soon reduced to a profitable level.

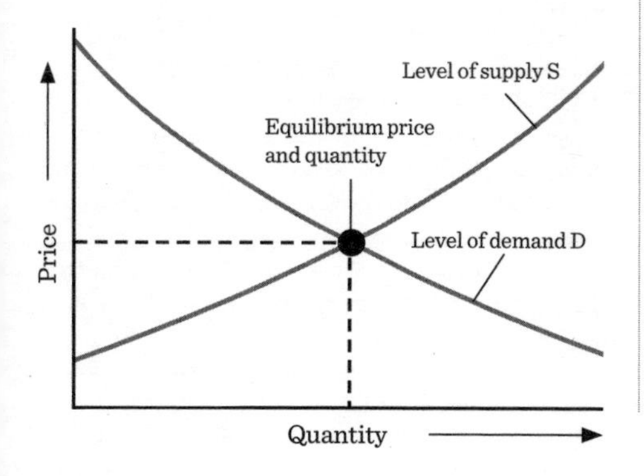

3 Are markets fair?

Economists worry a lot about a concept called 'Pareto efficiency'. This is an attempt to model how goods can be distributed between different parties in an optimal way, so that each person's 'utility' (basically, the enjoyment or benefit they get from their share of the goods) is maximized. Pareto improvements redistribute goods in the best way to improve utility for some without reducing it for others, until the most efficient allocation is reached. Economists since Adam Smith have tended to believe as an article of faith that trade in free markets can be used to create fair, Pareto-efficient outcomes.

4 Can markets be universally applied?

So-called 'welfare theorems' attempt to describe conditions in which markets produce Pareto-efficient outcomes, but while they may work for a limited set of idealized interactions, they often fall apart in real world situations. For example, the 'tragedy of the commons' is a case where everyone can freely make use of a common resource (for example water, grazing land or fish in a local pond), but their use depletes the amount available for everyone else. More challenges arise from 'public goods' – resources from which no one can be excluded, but which must be paid for somehow (such as street lighting). In such situations, regulation or taxation are the most logical solutions; market-based mechanisms often end up over-engineered and inefficient.

5 Are markets efficient?

Another fondly held belief for most economists is that markets are 'efficient'; in other words, the price at which goods change hands at any moment is both fair and rationally justifiable because all parties involved have a full understanding of all the factors involved and are acting in a sensible way that will maximize their utility. This assumption (the 'efficient markets hypothesis') is necessary in order for economists to make any useful observations or predictions, but it seems naïve to say the least; markets behave in irrational ways all the time, with greed and fear leading to herding behaviour that causes everything from minute-by-minute fluctuations in the price of stocks and shares to enormous asset price bubbles that culminate in catastrophic crashes. Behavioural economics is a relatively young field that attempts to take account of these human factors.

6 Inflation

When it comes to macroeconomics, one of the key concerns for most governments and central banks is inflation, the rate at which overall prices are rising (or, rarely, falling) over time. A slow but steady increase is, perhaps surprisingly, seen as a good thing; overall inflation is linked to the amount of money circulating in an economy, so when the economy is growing healthily and there's an excess of money, prices will rise to absorb it. Inflation erodes the value of 'unproductive' money left sitting in the bank, so it's supposed to encourage useful investment, and for governments it also has the benefit of diminishing the real value of debt on the national balance sheet, making it easier to pay back. Prolonged low inflation or deflation is, therefore, generally considered a bad thing, but equally so is excessively high inflation.

7 Fiscal and monetary policy

Governments have traditionally used two tools to regulate growth and inflation. Fiscal policy is taxation and public spending: the idea is that raising taxes can take excess money out of the economy, lowering inflation and providing funds for government to spend on useful projects (which may themselves encourage growth and raise further revenues , the 'fiscal multiplier' effect). Monetary policy, meanwhile, involves controlling the money supply by adjusting central bank interest rates (which underlie the rates offered by private banks). Higher rates encourage savers to keep money in the bank, discourage demand for new lending and generally reduce the amount of free money (and hence inflation), while lower rates flush unproductive money out of savings and boost demand for lending, encouraging consumption and economic growth while boosting inflation.

8 Exchange rates

Further complicating the economic picture is the exchange rate between currencies. Countries essentially have two options here; either to 'peg' the rate so that it's always worth a certain amount of another currency, or to let the value float free in currency markets. Pegged currencies give stability but remove many of a government's options for economic management, while free-floating currencies give a country full access to monetary policy but less control of prices on imported goods. Exchange rates can be manipulated to some extent through interest rates; a higher rate will tend to draw in investors and strengthen the currency, while a lower one will drive them off in search of better returns elsewhere.

9 Unemployment

The rate of unemployment is another crucial factor in economic management. Most economists agree that, though harsh on individuals, a certain level of unemployment is needed for an economy to run smoothly (if it falls too low, workers can argue for large pay rises that trigger inflation). Modern thinking pinpoints three distinct causes. Frictional unemployment is caused by people spending time between jobs, while structural unemployment is due to governmental issues, such as a lack of suitable skills for workers, or even a benefits system that disincentivizes job seekers (though in practice this is very rare). The third cause is a shortage of demand at certain times in a country's economic cycle; in theory, this should cause prices to fall and carry wages with them, but in practice companies either ignore the pay problem until they hit financial difficulties, or they solve the oversupply issue by laying off workers.

10 Growth theories

Economists define growth as an increase in 'real' gross domestic product (GDP) – that is, an increase in the value of all goods and services created in a country once inflation is taken into account. They separate long-run growth trends from the short-run fluctuations of the so-called 'business cycle' (a few years long). Long-run growth is widely agreed to be driven by technological change and improvements in productivity (each worker's contribution to the GDP). Explanations for the short-run cycle, however, vary; neoliberal economists tend to blame 'exogenous' factors (disruptions to a truly free market by the state and other players). Others, however, see dips in the cycle as 'endogenous' or market-based, perhaps even actively *requiring* government intervention to correct.

TALK LIKE A GENIUS

‘It’s tempting to imagine there’s a middle ground when it comes to exchange rates: for instance, allowing a currency to float within a fixed band. But attempts to do that usually end up being disrupted by currency traders who take advantage of being able to force the government’s hand to support its policy, and the results aren’t pretty. The UK had a famously bad experience with exchange rate bands in 1992, and even now the politicians still call the day it all crashed down “Black Wednesday”.’

‘Although the tax-and-spend of fiscal policy is often depicted as something that’s principally there for government to fund services, most economists would say that its main importance is really as a way of managing the economy. In fact, there’s a hard core that would say you don’t need tax to fund government spending at all; central banks can just conjure up more fiat money into existence with a tap of a keyboard. The catch is that extra money can cause inflation and devalue the currency already in circulation, so there are limited situations where you’d want to use it.’

WERE YOU A GENIUS?

1 FALSE – the relationship is the other way around, with prices rising in response to rising demand, and falling when demand drops.

2 TRUE – the benefit parties gain from different goods is described as their ‘utility’.

3 TRUE – although behavioural economists, at least, accept that this is rarely the reality.

4 TRUE – near-full employment is thought to give workers too much power to demand inflation-boosting pay rises.

5 TRUE – this is why interest rates and money supply can be used to influence inflation.

In theory, markets are a rational and efficient way of establishing price through supply and demand, but an overall economy includes many other complex factors.

Keynesianism and monetarism

'One of the great mistakes is to judge policies and programs by their intentions rather than their results.'

MILTON FRIEDMAN

Two contrasting approaches to economic management have shaped the modern world since the end of World War II. The ideas of John Maynard Keynes emerged from his *General Theory of Employment, Interest and Money* published in 1936, and dominated the era of postwar consensus. More recently, economic policy has been driven by the 'monetarist' ideas of Friedrich Hayek, Milton Friedman and the so-called 'Chicago School' economists. The faltering recovery from the 2008 economic crisis, however, saw renewed interest in Keynesian ideas.

Tax, spend and borrow, or manipulate the money supply? These are the two big ideas of postwar economics.

1 John Maynard Keynes argued that the amount of demand naturally adjusts to the amount of production in an economy.

TRUE / FALSE

2 Franklin D. Roosevelt's New Deal was a deliberate implementation of Keynesian economic policy.

TRUE / FALSE

3 The World Bank was formed after the breakdown of the Bretton Woods agreement in the 1970s.

TRUE / FALSE

4 Monetarists believe that the best way to raise inflation is by printing money to boost the money supply.

TRUE / FALSE

5 Supply-side economists argue that tax cuts should be targeted at the wealthiest in society since they are most likely to invest the money saved in new wealth-generating projects.

TRUE / FALSE

TEN THINGS A GENIUS KNOWS

1 Classical and neoclassical economics

Up until the turn of the 20th century, economic thought (aside from Marx's radical interpretation) followed a single school now known as 'classical economics'. Incorporating the ideas of Adam Smith, David Ricardo, Thomas Malthus and John Stuart Mill, it was liberal in its political outlook, arguing that markets were better able to regulate themselves with minimal state intervention, and that free markets and international trade were more beneficial than protectionism. Around 1900, a major breakthrough occurred with the realization by Englishman Alfred Marshall and various economists of the so-called 'Austrian School' that the labour theory of value was wrong; rather than the value of goods being determined by the labour put into them, the *price* was determined by laws of supply and demand. This new thinking defines neoclassical economics.

2 Say's Law

A deceptively simple economic law introduced in 1803 by French economist Jean-Baptiste Say states that aggregate production is always balanced by aggregate demand. In other words, if a country's overall economic output increases, the money generated by that will naturally be spent on goods and services. Reading between the lines, this means that if there's a lack of demand for one particular good, there must be an unmet demand elsewhere in the economy; market forces should naturally shift resources such as workers away from those areas suffering lack of demand and towards those with clamouring customers. Say's Law became a cornerstone of *laissez-faire* economics (an outlook that advocated minimal intervention in the market), since it appeared to show that a growing economy would, in the long run, tend towards full employment and maximize its efficiency without government intervention.

3 The Great Depression

In October 1929, the US stock market crashed after a decade of strong and steady growth and a runaway year that had inflated a huge asset bubble. Speculators and ordinary investors were left marooned; consumer spending plunged and could not be revived by the accepted neoclassical policy of lowering interest rates. Spending dropped still further as people began to anticipate a deflationary period of lower prices to come. As industry began to suffer, tariffs were introduced that, with hindsight, only worsened the situation. Most problematically of all, because countries were locked to the Gold Standard (the idea that their currencies should always be worth a certain amount of gold), they could be forced by speculators to mimic US policy in order to maintain the relative value of their currencies. The effects worsened through the early 1930s as recession deepened to depression; economies shrank worldwide and many millions of jobs were lost.

4 Keynes' *General Theory*

Faced with the Great Depression, English economist John Maynard Keynes argued for a radical solution: governments should inject money into economies through massive spending (financed by borrowing at low interest rates). Neoclassical economists resisted this view, arguing that Say's Law would eventually cause the economy to correct itself. The USA, however, was soon inadvertently putting Keynesian policy into action; faced with the prospect of complete social collapse, President Franklin D. Roosevelt launched his 'New Deal' in 1933, a mix of social guarantees and massive public spending that soon showed positive results. Keynes' 1936 opus *The General Theory of Employment, Interest and Money* set out to undermine the intellectual arguments for Say's Law; it might work in the long run, but as he pithily pointed out 'in the long run we are all dead'. The heart of the Keynesian argument is that over shorter timescales, it is aggregate *demand*, rather than supply, that decides the level of economic activity,

5 Bretton Woods

In 1944, with World War II still raging, delegates from 44 allied nations gathered at Bretton Woods, New Hampshire, to discuss how they would rebuild the postwar world. The meeting resulted in a monetary system that would eventually encompass the entire developed world outside the communist bloc, embracing the idea of free trade while seeking to control the speculative capital that had done so much to damage the prewar system. Keynes led the British delegation, and the Bretton Woods regime is often depicted as a Keynesian creation, but in reality many of his ideas for automatic stabilization and promotion of economic growth were rejected in favour of a more conservative system promoted by the United States.

6 The end of Bretton Woods

Under Bretton Woods, stability was ensured by a system of fixed exchange rates pegged to the dollar, which itself was convertible to gold. Exchange rates were administered by the newly established IMF, but the system created huge trade imbalances as the US economy grew through an export boom. Fixed rates limited the scope for economic management to fiscal policy (taxation, borrowing and spending), with little place for monetary tools. As pressures grew, countries started to abandon the system and let their currencies float. The USA, meanwhile, began to increase its own money supply and the dollar came to be seen as overvalued. President Nixon's 1971 declaration that the dollar would no longer be convertible to gold, effectively marked the end of the Bretton Woods system. In the following years, most other currencies abandoned their pegs to the dollar and became free floating.

7 The Chicago School

The end of Bretton Woods was (somewhat unfairly) seen as a rebuff for the Keynesian idea of economic management, but at the same time the relaxation of exchange rates paved the way for a new approach championed by economists at the University of Chicago. Friedrich Hayek, their mentor, was an Austrian émigré and a fierce advocate of neoclassical *laissez-faire* economics and what would today be described as libertarianism. Milton Friedman, his chief disciple, made some significant advances that appeared to undermine aspects of Keynesian theory. Together, they put forward an argument that the best means of tackling inflation and managing an economy was through monetary, rather than fiscal, policy.

8 Monetarism

The monetarist approach to economics, popularized under Ronald Reagan and Margaret Thatcher in the 1980s and dominant for almost three decades, argues that all the important aspects of an economy can be managed through adjustments to central bank interest rates. Lowering rates increases the money supply and promotes growth and inflation, while raising them encourages savings, lowers inflation and cools growth. These effects are hard to argue with, but politicians and economists with an ideological commitment to a minimal state, took things a significant step further, arguing that fiscal policy (direct government intervention via taxation and spending) was actively harmful.

9 Supply-side economics

Developing in parallel with monetarism, supply-side economics essentially argues that Say's Law, the idea of increased supply creating its own demand, was right all along. Its followers argue for low taxes, looser regulation, and privatization of 'inefficient' state-owned enterprise; the theory is that such reforms release more capital for investment and increase business growth to the benefit of all. Supply-side economists are particularly keen to cut taxes for the richest in society, arguing that these people are 'wealth creators' more likely to invest and create better jobs for the lower paid; hence, the approach has also been called 'trickle-down economics'.

10 Debt and austerity

Events after the financial crisis of 2008 threw a spotlight on the question of how much debt a government could sustain, and how it is best reduced. Debt spiralled during the crisis as governments were forced to spend large sums buying assets (including entire banks) in order to save the system from complete collapse, just as a crash in output and consumer confidence led to a contraction in taxation revenues. In the aftermath, countries adopted two broad approaches to the debt problem. Some borrowed further to finance a 'fiscal stimulus' of growth-boosting infrastructure projects and tax cuts for low earners (who are more likely to spend additional money). Others pursued fiscal consolidation or austerity, cutting government spending in a way that attempts to balance the annual government accounts, even if it does little to promote growth.

TALK LIKE A GENIUS

' I wonder if Keynes was consciously copying Einstein when he called it his *General Theory*? He was certainly setting out with the same idea of overthrowing a previous classical model... '

' The Laffer curve is a famous graph that seems to show how higher rates of taxation actually reduce government revenue. But it still raises some big questions; what's the ideal tax rate to maximize a government's income, and how high is too high? Ask two economists and you'll probably get a dozen different answers. '

' The different ways that governments tackled debt after the 2008 crash really emerged from different ways of looking at the entire economy. The austerity approach treats government borrowing like household debt – if you pile more onto the outstanding amount, you need to find a way of cutting your spending so you can start paying it down. The fiscal stimulus approach, on the other hand, basically says that if you can borrow cheap money and know somewhere worthwhile to invest it (a big new infrastructure project, for instance), then the fiscal multipliers mean you get more back in the long term, and don't have to cut your day-to-day spending so much in the short term. '

WERE YOU A GENIUS?

1 FALSE – Keynes believed that production (supply) adapted to meet fluctuating levels of demand.

2 FALSE – at least, Roosevelt never publicly acknowledged that he was implementing Keynesian policies.

3 FALSE – the World Bank was actually established at Bretton Woods.

4 FALSE – monetarists would always rather adjust the money supply through interest rates rather than printing new money.

5 TRUE – this is the basis of 'trickle-down' economics.

Keynesians argue that demand, rather than supply, is the force that determines economic growth. Monetarists tend to believe it's the other way round.

Postcapitalism

'[Capitalism] is not intelligent, it is not beautiful, it is not just, it is not virtuous – and it doesn't deliver the goods ... But when we wonder what to put in its place, we are extremely perplexed.'

JOHN MAYNARD KEYNES

The long aftermath of the 2008 financial crisis saw economies around the world endure a slow and faltering recovery. Some claimed this reflected a repeat of mistakes endured in the Great Depression and called for a return to Keynesian policies, but others questioned whether this wasn't an early sign of a more fatal malaise. Could capitalism, after enduring for several centuries, be entering its last days? And if so, then what might take its place?

Can we find a better way to better share the benefits of capitalism, or should we be contemplating lives with less work, perhaps under a very different system?

1 French economist Thomas Piketty has shown that return on capital typically outpaces the growth of the economy in general.

TRUE / FALSE

2 Economists generally assume that growth is driven by the discovery of new natural resources to exploit.

TRUE / FALSE

3 Workers' pension funds often pressurize companies into cutting their current workforces.

TRUE / FALSE

4 Market socialists argue that governments should take control of key elements of production.

TRUE / FALSE

5 Bill Gates and Elon Musk have both voiced concerns about the idea of a universal basic income.

TRUE / FALSE

TEN THINGS A GENIUS KNOWS

1 The downfall of capitalism

People have been predicting the doom of capitalism for centuries – since at least the writings of Karl Marx in the 1860s and of H.G. Wells a few decades later. Marx predicted that tensions between labourers and owners of capital would inevitably boil over into outright revolution, while Wells had a more utopian vision in which humanity would eventually evolve beyond such concerns. More recently, the influential *Limits to Growth* report of 1972 argued that limited availability of natural resources, coupled with issues such as pollution, put an upper limit on the size to which the global economy can grow. Since the concept of growth has been at the centre of global economics since the 18th century, this, too, is tantamount to saying that capitalism, as we know it, is on borrowed time.

2 The 'end of history'

While the 20th century did see major workers' revolutions, they didn't occur in the kind of economies that Marx predicted; instead of advanced bourgeois industrialized nations going communist, the two biggest examples were a society at the very beginnings of industrial growth (Soviet Russia) and one that was still mostly agrarian (China). Followers of Marx argue that this explains the troubles both revolutions experienced, with China turning to a system of market economics in just a few decades and the Soviet bloc collapsing around 1990 after its failure to deliver promised benefits. Nevertheless, Western democracies were not shy in proclaiming a victory for capitalism, with American political scientist Francis Fukuyama famously proclaiming it the 'end of history'; from here on out, neoliberal economics would reign unchallenged.

3 Reasons for the 2008 crisis

Those predictions of everlasting neoliberal hegemony lasted less than two decades before the cracks became dramatically apparent. It's very hard to argue that the causes of the 2008 global financial crisis were not fundamentally rooted in the *laissez-faire* assumptions of neoliberal economics and the policies this encouraged to create a seemingly endless illusion of economic growth. The spark came from the relaxation of US mortgage markets (in an era of low interest rates deliberately aimed at engineering a housing asset bubble), and it spread around the world thanks to a top-heavy superstructure of complex investment and trading products overlaid on the real economy. When the shaky nature of the US 'sub-prime' mortgages became clear, much of the panic that spread around the world was simply due to financial institutions not knowing how exposed they, or each other, were to the consequences.

4 Solutions to financial crises

The 2008 crisis was brought to an end not through monetarist remedies (with interest rates already low, scope to adjust them further was severely limited), but by concerted fiscal policy, as governments stepped in to buy large stakes in banks and inject money into the system. In the process, they piled on huge amounts of national debt, which the now-secured financial system then set about punishing them for. Now governments that attempted to continue spending in order to restart growth saw their credit ratings (decided by the very ratings agencies that had failed to see the coming crisis) threatened, and some were forced instead to pursue austerity policies of public spending cuts and reduced investment in an effort to reduce their debt (or at least prevent more from piling up). Meanwhile, despite low interest rates, businesses proved reluctant to make the capital investments that improve productivity, leaving growth anaemic at best.

5 *Capital in the 21st Century*

In 2013, French economist Thomas Piketty published his landmark study of wealth distribution over the past 250 years, sparking a debate that still has a long way to run. Piketty's central finding was that growing inequality is a built-in feature of the capitalist system that can only be countered by government intervention. The issue is that the return for those with capital to invest (not just business profits but also dividends, interest and income from rented property) typically exceeds the rate of economic growth that everyone else relies on for pay rises, benefits and the like. Thus, capital tends to accumulate and the rich get richer, while everyone else stays in the same place or gets poorer, an issue that is particularly acute in periods of low-to-no economic growth.

6 Inequality and politics

The question of what to do about rising inequality has become a huge one, exacerbated by the way globalization has hit the security of the working class in developed nations. Often the neoliberal argument depicts these huge accumulations as the just rewards of capitalism, from which we will all eventually benefit through trickle-down economics. Elsewhere, however, howls of outrage against 'the 1 per cent' are commonplace (though it's really the 0.1 per cent we have to worry about, as the problem identified by Piketty gets more extreme the higher up the ladder you go). Piketty's proposed solution is some sort of wealth tax, but such a tax would have to be very carefully designed, and would require unprecedented international cooperation in order to implement.

7 The threat of automation

Meanwhile, another approaching threat seems likely to force change, whether governments want it or not. Automation has already done much to reduce the industrial workforce of advanced nations, where it's cheaper, faster and safer for robots to do many jobs. Workers have so far found new employment (if not actual careers) doing the sort of jobs that only humans can (such as driving), but advances in artificial intelligence are now threatening those, as well as a whole swathe of middle-class jobs that rely on, for example, reading and interpretation of documents. Once again, the blind motive of capital to increase productivity and profit is responsible, but what happens if the number of jobs that require actual human input falls dramatically?

8 Universal income and other solutions

One thing that is certain is that businesses will still need consumers – a population reduced to subsistence on meagre benefits will drag the capitalist system down with it. So if there aren't enough well-paid jobs to go round, what's the solution? One widely touted idea is universal basic income or UBI. This is an amount, sufficient to cover basic living expenses, paid to every adult citizen in a country without condition, replacing most other benefits. Advocates argue that it could be paid for through taxes on additional income earned from jobs, through a wealth tax, or by taxing businesses on the basis of how many jobs they turn over to automation. Alternative proposals include a job guarantee (where the public sector acts as an employer of last resort), and enforced private-sector employee quotas.

9 The end of growth?

A key question, however, is whether economic growth will continue to be sustainable for all countries in the medium-term. Some economists suggest that the answer to that question is no, and that a crisis will emerge from within as the rapid growth of capital overwhelms everything else. Others, however, think that even without such a crisis, we should be looking at alternative systems now. Some of the most radical ideas have emerged from the environmental movement, where economic thinkers are concerned that all growth is ultimately rooted in the exploitation of (ultimately limited) natural resources. Others simply argue that GDP and per-capita income are poor ways of measuring a nation's welfare; we should perhaps reduce our dependence on consumption and think about things like happiness instead.

10 What's the alternative?

Two of the leading postcapitalist approaches are degrowth and market socialism. Degrowth originated in the 1960s and argues for deliberate economic contraction and reduced consumption, while encouraging individuals to reap the rewards to be found in a simpler lifestyle, less stress and more leisure time. Market socialism, meanwhile, originates in the thinking of liberal economist John Stuart Mill. Late in his career, Mill argued that while retaining the broad operations of a market economy, it should nevertheless be possible to replace capitalist businesses with workers co-operatives to ensure a more equitable distribution of the rewards. Modern market socialism supports a variety of other forms of profit-redistribution, including state ownership and social enterprises.

TALK LIKE A GENIUS

‘The problem with austerity is that it's a vicious circle – every government cut also tends to reduce government income and take money out of the economy, which doesn't do anything to encourage growth.’

‘Some people have said that Piketty's *Capital in the 21st Century* takes 696 pages just to prove that inequality increases when *r* (the return on capital) is greater than *g* (economic growth), which is most of the time. But at least he includes a lot of evidence to support that.’

‘The most strident criticism of UBI is that it would act as a disincentive to work, but that may not be a problem if there simply aren't enough jobs to go around. Nevertheless, the idea has attracted surprising support from across the political spectrum: the libertarian right like it because it cuts government bureaucracy, the left because it raises people out of poverty.’

‘If we ever get a UBI, it might make people rethink the value of work; a lot of people would be able to do things they love and try to better themselves, and boring jobs would have to pay decent wages to get people interested.’

WERE YOU A GENIUS?

1 TRUE – as a result, overall capital in the hands of the wealthy tends to grow faster than any other part of the economy.

2 FALSE – most economists believe that growth is driven by technological advances that improve productivity.

3 TRUE – a company's pension fund is often one of its major shareholders, but the fund is driven to increase profits, even if that means laying off employees.

4 FALSE – market socialism argues for various forms of shared ownership against a free-market background.

5 FALSE – Musk is believed to support the idea, while Gates has argued that it could be funded via a ‘robot tax’.

Capitalism still hasn't really recovered from the crisis of 2008, and some argue that developments in the near future mean it will have to change or die.

Environment and climate change

'[One year] doesn't make a trend, but this does: 14 of the 15 warmest years on record have all fallen in the first 15 years of this century.'

PRESIDENT BARACK OBAMA

Two hundred and fifty years of industrialization have triggered environmental changes of a kind normally seen only on geological timescales, but it's only in the past few decades that we've fully recognized the problem. The issues facing our planet today are a concern not only for environmental pressure groups, but also for politicians and economists seeking to shape the future of society.

The future of our planet hangs in the balance – what are we going to do about it?

1 Climate scientists can calculate past global temperatures by measuring the height of trees.

TRUE / FALSE

2 Atmospheric carbon dioxide levels have risen by almost 25 per cent since the Industrial Revolution.

TRUE / FALSE

3 Under the Paris Agreement on climate change, countries will work to limit the global average temperature rise to 1.5 °C (2.7 °F) above pre-industrial levels.

TRUE / FALSE

4 Forty years ago, some climate scientists were predicting that Earth was heading for a new Ice Age.

TRUE / FALSE

5 The Gaia hypothesis means that the Earth's geological and biological systems will naturally counteract global warming without our intervention.

TRUE / FALSE

TEN THINGS A GENIUS KNOWS

1 Earth's changing environment

The idea that the Earth can go through long-term changes in its environmental conditions emerged through the discoveries of palaeontologists and geologists in the 18th and 19th century. Scotland's James Hutton, widely considered as the founding father of modern geology, suggested that glaciers had once been widespread in the northern hemisphere in the second edition of his *Theory of the Earth* (1795). This idea was later put on a firmer geological footing thanks to the work of Swiss-American Louis Agassiz, who named the event the 'Ice Age'. In the 1840s another Scot, Charles Lyell, traced the origin of the coal being exploited by new industries in Europe and America to ancient tropical forests that had once been widespread even at high latitude.

2 Greenhouse gases

The question of what could cause such major environmental changes, however, remained elusive. French physicist Joseph Fourier had made an important first step as early as 1824, when he showed that Earth's atmosphere keeps our planet warmer than it would otherwise be. In the 1850s, Eunice Newton Foote in America and John Tyndall in Britain both independently discovered that carbon dioxide exercised a particularly strong warming effect by trapping sunlight reflected from surfaces (such as the surface of the Earth). Tyndall also identified another significant 'greenhouse gas' in the form of methane. Some geologists began to wonder whether past climate changes might be linked to changes in the amounts of these gases present in the atmosphere driven by variations in volcanic activity.

3 The Gaia hypothesis

In 1896, Swedish chemist and physicist Svante Arrhenius assembled many pieces of the climate jigsaw for the first time. He not only put numbers on the way that changing concentrations of atmospheric carbon dioxide would affect average temperatures on Earth, but also noted an array of other influences. For instance, rising temperatures would cause levels of snow and ice cover to diminish, lowering Earths' ability to reflect solar radiation back into space and increasing the warming effect. In the 20th century, geoscientists found many other feedback mechanisms of this sort; some accelerate changes once they start, while others slow them down. Ultimately, this led to the Gaia hypothesis, a still-controversial but nevertheless powerful model that envisages Earth and its living biosphere as a unified system that is at least partly capable of self-regulation.

4 Sources of climate change

Even in the late 19th century, Arrhenius and others realised that human industry was releasing vast amounts of carbon into the atmosphere. Earth's present-day plant life (which absorbs carbon dioxide in order to grow) stood little chance of counteracting the effects of rapidly burning millions of years' worth of fossil fuels, so a rise in carbon dioxide levels was inevitable. What they underestimated, however, was the speed of that release; Arrhenius thought that the planet would slowly warm over thousands of years. Furthermore, the link between carbon and climate was lost amid early 20th-century debates about the influence of solar activity, and Serbian scientist Milutin Milankovitch's identification of cyclical variations in Earth's orbit that change the amount of sunlight it receives and create long-term climate oscillations (such as those seen during ice ages).

5 The anthropogenic effect

The first person to suggest that anthropogenic (human-induced) climate change was already underway was British engineer Guy Stewart Callendar, who in 1938 produced evidence for a rise in both carbon dioxide and global temperature over the preceding 50 years. However, many scientists ignored Callendar's findings, or assumed that self-regulating mechanisms would take care of the problem. It was only in the 1950s that atmospheric and oceanographic research showed that some of the natural 'sinks' predicted to absorb rising carbon dioxide were not as effective as previously thought. In 1960, US scientist Charles David Keeling compiled global data showing the rising trend in carbon dioxide – a graph known as the Keeling Curve that continues to show an inexorable rise to the present day.

6 Rising temperatures

Ironically, a major barrier to recognizing the long-term global warming trend arose from another human pollutant – aerosols. Tiny particles released from pressurized spray cans and other sources have a significant cooling effect in the atmosphere, and as their use spread after World War II, global temperatures actually cooled on average. It was only when other problems arising from aerosols (chemical smogs and the hole they created in Earth's protective ozone layer) forced action to reduce aerosol use, and levels began to diminish in the 1980s, that the effects of carbon dioxide reasserted themselves. Since then, the trend has become clear – global temperatures and carbon dioxide levels are rising together.

7 There is no 'debate'

An estimated 97 per cent of environmental and climate scientists working today agree that global warming is both real, and anthropogenic in its origin. No respected scientific organization argues against the idea, but still many politicians and commentators seem to treat the question as open to debate. In part, this is down to a misunderstanding of terms and our own cognitive biases; we tend to privilege our personal experience of short-term, localized weather over the more abstract graphs showing long-term global trends in climate. What's more, because climate scientists know Earth's weather is a hugely complex and sensitive system, they unwittingly create mixed messages by refusing to link specific weather events to global climate trends. When the media promotes a few genuine contrarians and a larger number of pressure groups for vested interests to equal footing with the entire scientific establishment in the interests of 'balance', it's a recipe for further confusion.

8 Environment and economics

From an economist's point of view, the environment is an extreme example of a 'public good', a special case in which not only national but global government intervention is necessary for the benefit of all. The United Nations Intergovernmental Panel on Climate Change (IPCC) was established in 1988 and although the road has been far from smooth, progress towards an agreement on tackling the problem or at least limiting its worst effects has been made with the Kyoto Protocol of 1997 and the Paris Agreement of 2015. Sceptics argue that limiting carbon emissions will harm the economy, but in doing so they overlook new opportunities. Most economic growth theories agree that technological advances are a key driver of prosperity; properly handled , the transition to renewable energy can provide a huge economic boost as well as saving the environment.

9 Market and government solutions

A major challenge, however, lies in the unwillingness of modern governments to make direct interventions in the free market. Instead, most seek to incentivize good behaviour by tipping the scales rather than introducing outright limits on carbon production. One such example has been the creation of markets in 'carbon credits', allowing companies to effectively trade in the 'right to pollute' (with environmentally friendly companies selling unused credits to heavy polluters). The theory is that the quantity of carbon credits in the market can be reduced over time to bring down the overall level of emissions, but whether this is really better than more direct forms of regulation is a matter of opinion.

10 Geoengineering and carbon capture

While most efforts concentrate on deploying renewable energy schemes and limiting fossil fuel use, is there any hope of reversing the damage that's already been done? Various ambitious geoengineering schemes aim to do just that, though plans to artificially cool the Earth using mirrors in space or by flinging particulates into the upper atmosphere raise concern about making further alterations to a system we still barely understand. If there is to be a 'way back', it more likely lies in restoring the atmosphere's natural balance through withdrawal of carbon dioxide. Carbon-capture schemes range from large-scale forestry to the use of chemical filters, often at the production end of carbon-emitting industrial processes. Whether they can offer an economic solution without further disrupting our planet's delicate environment, however, remains to be seen.

TALK LIKE A GENIUS

‘Modern environmentalism began with two events in the 1960s. Rachel Carson’s 1962 book *Silent Spring* linked the widespread use of pesticides to declining bird populations in parts of America, and showed how much we had to learn about the complexity of the natural world. Then Apollo 8’s flight around the Moon in December 1968 took people far enough away from Earth to see the planet as a whole for the first time. Billions tuned in to the broadcasts and got a wake-up call as to how small and fragile the Earth really is.’

‘Climate change might be easy to ignore for those of us in countries with temperate climates, but it’s already causing rising sea levels, famine and drought in other parts of the world. Left unchecked, it could render parts of the planet effectively uninhabitable, creating problems that will inevitably come home to roost for developed nations.’

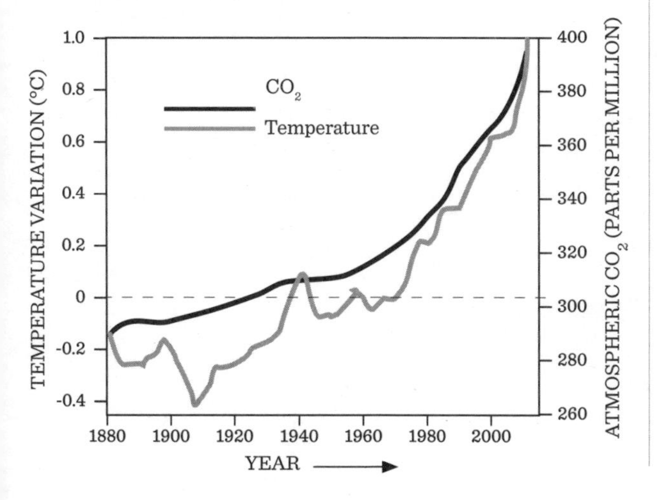

WERE YOU A GENIUS?

1 FALSE – scientists work out temperatures by looking at their changing width (in the form of annual growth rings inside tree trunks) among other sources.

2 FALSE – they’ve actually risen by more than 40 per cent.

3 TRUE – and they will do their best to keep the rise to just 1 °C (1.8 °F).

4 TRUE – however this was to do with the discovery of much longer-term climate cycles which some popular books then exaggerated into an imminent threat.

5 FALSE – Gaia suggests Earth’s climate can adjust to correct for small long-term changes, but not relatively sudden and dramatic ones.

All the evidence says that climate change caused by industrial greenhouse gases is a critical problem for all life on Earth – and we need to start repairing the damage right now.

Fermat's Last Theorem

'I don't believe Fermat had a proof. I think he fooled himself into thinking he had a proof.'

ANDREW WILES

Perhaps the most famous problem in the history of mathematics was finally solved (after more than three and a half centuries) by mathematician Andrew Wiles in the mid-1990s. Pierre de Fermat's theorem is deceptively simple to state, but fiendishly difficult to prove – indeed, doing so required Wiles to venture into realms of mathematics that did not even exist in Fermat's time. However, the long struggle to prove the theorem tells us a lot about mathematics, numbers and the wider principles of mathematical proof – all topics that any genius should know about.

Beautiful but ultimately useless, Fermat's Last Theorem pushed mathematicians to the limit in search of proof.

1 Fermat's Last Theorem claims to answer a simple question about whole-number solutions to fairly basic equations.

TRUE / FALSE

2 A mathematical proof by contradiction aims to show that a hypothetical mathematical relationship is never contradicted by reality.

TRUE / FALSE

3 Mathematicians have shown that you can calculate the square roots of negative numbers.

TRUE / FALSE

4 A mathematical group contains numbers that all share certain characteristics.

TRUE / FALSE

5 A Diophantine equation is one that can be solved using methods laid out by the Greek mathematician Diophantus.

TRUE / FALSE

TEN THINGS A GENIUS KNOWS

1 Pythagoras's theorem

Fermat's Last Theorem is rooted in one of the few bits of schoolroom geometry that almost everyone remembers: Pythagoras's theorem. This is the simple statement that for a right-angled triangle, the square of the hypotenuse (the longest side of the triangle) is equal to the sum of the squares of the other two sides. In other words, if you drew three perfect squares out from the three sides of a triangle with a 90-degree corner, you'll find that the areas of the smaller two fit exactly inside the largest one. If we designate the edges (from shortest to longest) as a, b and c and remember that the area of a square is simply its length times its height (both of which are identical), we can write down the relationship as: $(a \times a) + (b \times b) = (c \times c)$, or $a^2 + b^2 = c^2$. This is a simple mathematical equation, a statement that the numerical value of two quantities is identical (the meaning of the '=' sign and the root of the word 'equation' itself).

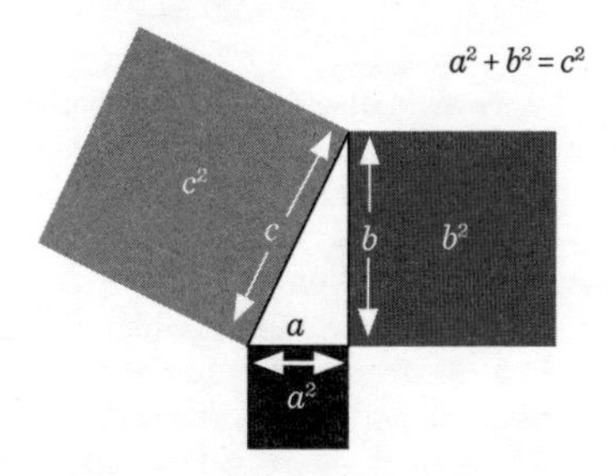

2 Pythagorean triples

One of the most interesting aspects of Pythagoras's theorem is that it has certain solutions in which a, b and c are all whole numbers or 'integers'. For example, if $a = 3$ and $b = 4$ then c must be 5 (because $3^2 + 4^2 = 9 + 16 = 25$, which is 5^2). Similarly if $a = 5$ and $b = 12$, then $c = 13$. Groups of numbers with this neat relationship are known as 'Pythagorean triples', and in theory there's an infinite number of them. As a geometric trick they have practical uses (long before Pythagoras, the ancient Egyptians are thought to have used the 3/4/5 triangle to establish right angles when building the pyramids), but shorn of their geometric meaning and treated solely as numbers, they are also intriguing from a purely mathematical point of view.

3 Diophantine equations

In the third century CE, Greek mathematician Diophantus of Alexandria made equations with integer solutions the centrepiece of a book called the *Arithmetica*. The equations he focused on are the typical stuff of school textbooks; statements that certain variables (these days symbolized by letters such as x, y and z) multiplied by themselves or by certain fixed constants (either written as numbers if known, or themselves symbolized a, b, c, etc.) balance with another combination of variables and constants. A simple Diophantine equation is $ax + by = 5$ (constant a times variable x + constant b times variable $y = 5$). The Pythagorean equation $a^2 + b^2 = c^2$ is another example. Although the *Arithmetica* was largely forgotten with the decline of the Roman Empire, parts of it were rediscovered and translated by Arab scholars in the 10th century. Here, the Diophantine method became known as *al-jabr* (the reunion of broken parts), the origin of our modern word 'algebra'.

4 Generalized solutions

Diophantus was interested not only in finding solutions to equations (the missing constant values that make them true for any value of the variables), but also in proving whether or not solutions existed at all. One key step in finding such proofs is to see if they can be generalized; in other words, can any actual numbers that appear in an equation be replaced by additional variables without breaking something? The root of Fermat's Last Theorem lies in the question as to whether you can generalize the Pythagorean equation; what if you replace the squares in $a^2 + b^2 = c^2$ with cubes ($a \times a \times a$, etc.), or powers of an arbitrary number n? Is there *any* value of n, other than 2, for which $a^n + b^n = c^n$ has whole-number solutions?

5 Fermat and his theorem

Pierre de Fermat was a French lawyer and amateur mathematician who made a huge variety of mathematical and physical discoveries during a long career. Like many mathematicians of the time, he was fascinated by the Diophantine equations, and much of his surviving correspondence revolves around proofs associated with them. Around 1637, Fermat wrote in the margin of his copy of *Arithmetica*, answering the question about generalizing the Pythagorean equation with an emphatic negative: 'It is impossible to separate a cube into two cubes, or a fourth power into two fourth powers, or in general, any power higher than the second, into two like powers. I have discovered a truly marvellous proof of this, which this margin is too narrow to contain.'

6 Did Fermat prove his theorem?

Fermat's immortal note was only discovered after his death, and the lack of any further mention from the man himself in his remaining three decades of life suggests that he soon realized he had made a mistake somewhere in his supposed general proof. The essential puzzle remained a hot topic, so it would be somewhat odd if Fermat kept a viable proof to himself. Nevertheless, he *did* discover a proof that there are no integer solutions to the equation for fourth powers (that is, $a^4 + b^4 = c^4$). This offers a useful insight into a common mathematical technique, the 'proof by contradiction'.

7 Proof by contradiction

Fermat's proof for fourth powers relies on considering the *opposite* situation to that being investigated, and showing that if such a situation were true, it would give rise to a logical contradiction of some kind. In this case, Fermat looked at the possibility that there *is* a solution to the equation $a^4 + b^4 = c^4$ with whole numbers, and showed that this was impossible. This is because if you assume that one solution (using arbitrary values) exists, it *always* implies the existence of another, smaller solution, and that would require an (impossible) *infinite* number of integers smaller than the arbitrary values. This specific type of contradiction proof is called an 'infinite descent'.

8 Imaginary numbers

Over the following centuries, mathematicians remained obsessed with Fermat's Last Theorem. Some pursued proofs for individual powers in the hope that they might hold the key to general solutions, while others hoped that proof by infinite descent might point the way (in other words, can you, by assuming that $a^n + b^n = c^n$ for some arbitrary value of n, reveal that the equation must also be true for a smaller value of n? In 1770, German Leonhard Euler used the concept of 'imaginary numbers' to prove that there are no cases where $a^3 + b^3 = c^3$. These are constructions such as the square root of –1, which are impossible to visualize in the real world (since multiplying a negative number by itself results in a positive number) but nevertheless mathematically powerful.

9 Number theory

While mathematicians continued to prove the theorem for certain specific cases, this was a long way from proving it for all possible powers, so others started taking a different approach. Around 1815, trailblazing French philosopher Sophie Germain did some of the first work on a general method using advances in 'number theory' (the study of the properties and relationships between different types of numbers) previously made by Carl Friedrich Gauss. In essence, she was able to show that the theorem has to be true for all numbers within a particular 'group' (an algebraic structure defined by the common properties of the objects within it, such as the natural 'counting' numbers), and then calculate which numbers are members of the group. This ultimately confirmed that the theorem was true for all exponents (values of n) up to 100; others used the same technique to expand proofs still further in the 20th century.

10 Andrew Wiles' solution

For most of the 20th century, however, mathematicians considered a general solution to Fermat's Last Theorem to be so hard as to be impossible without a fresh fundamental insight. Andrew Wiles' generalized proof is just that – a piece of genius-level lateral thinking linking apparently disparate areas of mathematics. In 1955, Japanese mathematicians Goro Shimura and Yutaka Taniyama had made the first step with what became known as the 'modularity theorem' – a proposed link between patterns in two independent 'groups' called modular forms and elliptic curves. Then in the mid-1980s, Germany's Gerhard Frey and American Ken Ribet found a link between modularity theorem and Fermat's theorem. Wiles' sideways approach to the problem, therefore, was to prove the modularity theorem rather than setting out to prove Fermat directly, and his success in 1994 brought a long chapter in the history of mathematics to a close.

TALK LIKE A GENIUS

❛ One of the lovely things about the theorem, of course, is that it's pretty useless – there was no great practical application waiting around the corner. Carl Friedrich Gauss, who knew a thing or two about maths, said he wasn't really interested as he could write down many similarly unprovable statements of no great import. If there's anything practical to be learned from the whole saga, it's probably in the way that widely separated areas of maths suddenly turn out to have deep connections. ❜

❛ A lot of the interest in the theorem was driven by fame and money. The French Academy of Sciences offered large rewards to anyone who could prove it in 1816 and 1850, and that was when the theorem really took off as a subject for study. In 1908, a German industrialist left 100,000 gold marks in his will to anyone who could publish an accepted proof in the next century, so Andrew Wiles just got in about a decade before the deadline. He also did well out of the Abel Prize (the Nobel of Maths), which he picked up in 2016. ❜

WERE YOU A GENIUS?

1 TRUE – this makes it all the more surprising that it took so long to prove.

2 FALSE – contradiction proofs actually show that the *opposite* case to the hypothesized relationship must always be false.

3 FALSE – the definitions of negative numbers and square roots mean that you can never get a meaningful numerical answer to the square root of a negative. But mathematicians can still do a lot of useful maths by treating them as 'imaginary' numbers.

4 TRUE – although the concept of groups is not purely limited to numbers, but can include other mathematical 'objects', such as geometric figures.

5 FALSE – a Diophantine equation is defined by its structure rather than its solution, and there are still some that seem to be unsolvable.

The concept of Pythagorean triples can't be extended from squares to higher powers – but it took 357 years to prove it.

Gödel's incompletness theorems

'Either mathematics is too big for the human mind, or the human mind is more than a machine.'

KURT GÖDEL

There's nothing mathematicians like more than an unsolved problem, but are they missing something fundamental about the nature of their pursuit? In 1931, German mathematician Kurt Gödel published a result that challenged the very foundations of not just maths, but many other sciences that rely on it. The incompleteness theorems don't generally affect our day-to-day use of mathematics and numbers, but they do raise philosophical questions about how much we can see mathematics as representing a fundamental 'truth'.

If you can't rely on the certainties of mathematics, what can you rely on?

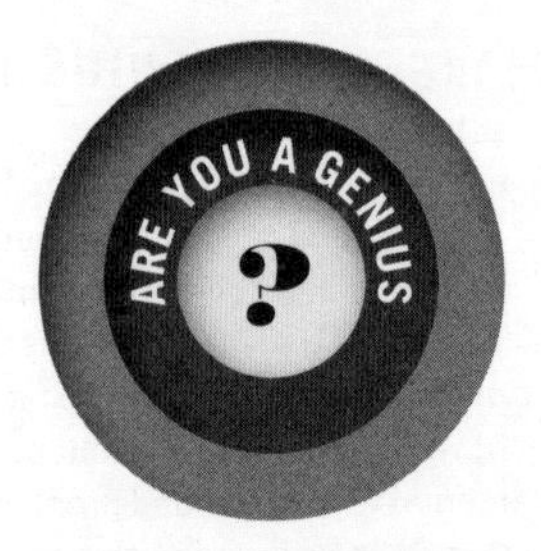

1 A mathematical axiom is a truth that reflects aspects of the real world beyond numbers.

TRUE / FALSE

2 Formalism is an approach to mathematics that aims to set down fundamental logical rules that represent objective reality.

TRUE / FALSE

3 Euclid's axioms are still the only way of describing the rules of geometry more than 2,000 years later.

TRUE / FALSE

4 Bertrand Russell and Alfred North Whitehead's *Principia Mathematica* is an attempt to prove that the truth of mathematics is ultimately dependent on the language used to describe it.

TRUE / FALSE

5 The incompleteness theorem shows that we can never have complete confidence in a mathematical system.

TRUE / FALSE

TEN THINGS A GENIUS KNOWS

1 Hilbert's second problem

Around 1900, German mathematician David Hilbert set the agenda for 20th-century maths with a list of no fewer than 23 major unsolved problems. Some of them are too abstruse to tackle here, and many have since been solved, but perhaps the most fundamental was his second problem: finding a proof that the 'axioms' of arithmetic are consistent. To break that down a little, the axioms of any theory are its basic premises, the rules that are assumed to be true within it. In order to be 'consistent', they must simply not give rise to any contradictions. Arithmetic, meanwhile, is the simple, familiar branch of mathematics involving addition, subtraction, multiplication and division of numbers.

2 Axiomatic systems

Axioms have always been at the heart of mathematics, giving rise to its power and underlining its historical status as a source of incontrovertible proof. The basic idea is that having established the rules of engagement, you can take a statement about a field of mathematics and assess whether or not it is true simply through logical application of the underlying axioms. The most famous set of axioms are the five outlined by Greek philosopher Euclid in his *Elements* (*c.* 300 BCE) in his description of geometry, but similar axioms, spoken or unspoken, underlie all branches of mathematics.

3 Logicism

The axioms of arithmetic that Hilbert wanted to test emerged from a project called 'logicism' that began with the work of German mathematician Hermann Grassmann. In 1861, Grassman pointed out that many of the principles of arithmetic and the properties of natural numbers, the positive whole numbers we use in counting and ordering, could be established using two simple mathematical operations. One of these was the successor function (loosely speaking, the addition of 1 to a previous number), while the second was the proof technique known as 'mathematical induction'. Inspired by this discovery, several mathematicians now attempted to define a complete set of axioms that could be used along with these techniques to define all of arithmetic.

4 Mathematical induction

As a common form of proof, induction is worth taking a moment to understand. Technically speaking, it's a 'direct proof', since it positively proves the theorem being investigated (as opposed to a proof by contradiction, which demonstrates that the theorem cannot be false). Induction involves two steps, first proving that the statement is true for a 'base case' (typically, the situation when a variable natural number n is 0 or 1). The so-called 'inductive step' is to then prove that if the statement is true for one natural number n, it must also be true for $n + 1$. So if the statement is true for one natural number, it is shown to be true for all.

5 Finding the axioms of arithmetic

Several mathematicians made independent attempts to find axioms that would describe all arithmetic in the late 19th century, including Charles Sanders Peirce in America and Richard Dedekind and Gottlob Frege in Germany. However, the most complete statement is generally agreed to be that formulated by Italian Giuseppe Peano in 1889. Peano's axioms first define 0 as a natural number (since otherwise its status might be in doubt). They then establish what equality means, such as the fact that it is 'symmetric' (if $x = y$ then $y = x$) and 'transitive' (so if $x = y$ and $y = z$ then $x = z$), and other fundamental rules. Finally, Peano describes addition and multiplication as mathematical functions that map two numbers (what we might call 'inputs') onto a third (the 'result'), for instance, confirming that the $x + y$ gives the same result as $y + x$. Much of this might seem like a statement of the obvious, but it's all necessary to give arithmetic a logical grounding.

6 Is maths a game?

Despite the determination of the logicists, however, doubts about the integrity of axioms were emerging elsewhere, particularly in the field of geometry. The possibility of 'non-Euclidean' spaces in which the normal rules do not apply (most critically, Euclid's fifth axiom that parallel lines never intersect) had troubled mathematicians for centuries, but from 1830 onwards, some began to develop complete descriptions of the axioms of such spaces. Euclid's own axioms, meanwhile, were found to be incomplete and to make too many assumptions (Hilbert himself drew up an improved system of 20 geometric axioms). The ability to describe systems using different sets of axioms spread concern that perhaps they were not a reflection of fundamental realities after all (as the logicists believed), but merely rules of a particular mathematical game. This attitude is known as 'formalism'.

7 *Principia Mathematica*

The ultimate attempt at a logicist description of mathematics was the monumental *Principia Mathematica* (1910–13) produced by Alfred North Whitehead and Bertrand Russell. This famously difficult book attempted to lay out a set of axioms and rules of inference against which all mathematical statements could, in principle, be tested for their truth. Mostly written in dense symbolic language with many cross-references and only brief interludes of English, the *Principia* famously takes 300 pages to prove the statement 1 + 1 = 2. It also has to go to great lengths to redefine apparently common-sense ideas, such as sets (collections of mathematical objects such as numbers or geometric figures), in order to avoid paradoxes.

8 The incompleteness theorems

Despite its apparent success, the *Principia* had some notable problems, and was soon realized to be incomplete – in other words, its axioms did not offer a full solution for every conceivable problem within the fields it set out to describe. In 1930, German mathematician Kurt Gödel showed exactly why this was the case with his first incompleteness theorem. This shows that, for any sufficiently complex mathematical system such as the *Principia*, there will always be statements that can be made using its axiomatic language, but which are unprovable – equivalents of asking someone to prove the statement 'This statement cannot be proved'. The following year, Gödel extended his thinking with a second theorem: no formal system of arithmetic can contain a provable statement of its own consistency. In other words, you can't prove the statement 'This system contains no contradictions' unless there are contradictions.

9 Implications of incompleteness

For mathematicians of the time, Gödel's theorems, once pointed out, were so self-evident that they brought the logicist program to a shuddering halt for some time (David Hilbert supposedly abandoned his attempts to improve on the *Principia* systems on the spot). However, the idea proved too useful to die completely and attempts to develop axiomatic descriptions soon resumed (albeit while acknowledging their limitations). The consequences of incompleteness, meanwhile, rippled out across philosophy and science. The theorems had deep implications for epistemology (the philosophical theory of knowledge, which had adopted Russell and Whitehead's work wholeheartedly to gain new insight). Specifically, they showed that statements could be true, despite not being provable.

10 Incompleteness and the brain

Some neuroscientists and philosophers argue that the incompleteness theorems can also provide insight into the nature of the human mind, and in particular the question of whether our brains are computers, capable of processing mathematical algorithms in the same way as an idealized model called a 'Turing machine'. If this is the case, then the brain would also be subject to Gödel's limitations and incapable of proving its own consistency. Most scientists and mathematicians who have engaged with this debate argue that human brains are indeed incapable of this task, so they can be modelled as machines. Some, however (notably British philosopher J.R. Lucas, but also Gödel himself), have argued that humans can recognize consistency unprovable by machines, with important implications for both the nature of our minds, and the question of whether we have free will.

TALK LIKE A GENIUS

‘ Gödel’s theorems might have persuaded Hilbert himself to give up on his project, but plenty of mathematicians think the Second Problem is still open. Some think you can use other forms of logic to get around them, while others think the incompleteness theorems might not apply to every possible system. ’

‘ Russell himself did a lot to open the whole incompleteness can of worms when he came up with something called “Russell’s paradox” in 1901. It’s a statement about set theory that says “Let R be the set of all sets that are *not* members of themselves”. So is R a member of itself, or not? It’s not directly connected to arithmetic, of course, but it did help undermine people’s confidence in finding complete axiomatic approaches. A more real-world, but less logically watertight version is the barber paradox – if the barber is the man who shaves all the men in a town who do not shave themselves, but *only* them, then does he also shave himself? ’

‘ If our brains are “just” machines, they’d have to be vulnerable to the incompleteness theorem, too. That means they can’t be both self-consistent and complete – though have you ever met a human brain that was either? ’

WERE YOU A GENIUS?

1 FALSE –axioms describe the rules of mathematical logic within certain systems, but don’t necessarily reflect any wider reality.

2 FALSE – it’s actually an approach that says mathematical systems can only be judged within the framework of language used to describe them.

3 FALSE – there are several other ways that are useful in specific situations.

4 FALSE – Russell and Whitehead actually wanted to prove that maths represents a fundamental truth independent of language.

5 TRUE – a system cannot be both complete and consistent, but that doesn’t mean it can’t be accurate in the vast majority of situations.

Axiomatic systems once seemed to offer a logical foundation for all of maths, but Kurt Gödel showed that they’re not foolproof.

The Riemann hypothesis and Goldbach's conjectures

'If I were to awaken after having slept for a thousand years, my first question would be: has the Riemann hypothesis been proven?'

DAVID HILBERT

Prime numbers are an endless source of fascination to mathematicians – though their definition is simple, the search for techniques to identify them without relying on brute-force calculation has lasted for at least two millennia. The Riemann hypothesis and Goldbach's conjecture are both statements about the nature of primes that appear true for all tests so far attempted, but have not yet been proved.

Primes are the building blocks of all numbers – but can we predict the way they're distributed?

1 A prime number is one that cannot be divided by any number other than 1 to produce a whole-number result.

TRUE / FALSE

2 The frequency of primes falls away in a smooth decline at higher numbers.

TRUE / FALSE

3 The Riemann hypothesis is a way of rapidly calculating whether large numbers are made up of two primes multiplied together.

TRUE / FALSE

4 Goldbach's strong conjecture says that any sufficiently large number is the sum of three primes.

TRUE / FALSE

5 The Riemann hypothesis has been proved to hold true in up to ten trillion individual cases, but nevertheless remains unproven as a general rule.

TRUE / FALSE

TEN THINGS A GENIUS KNOWS

1 Prime numbers

By definition, a prime is a positive natural number greater than 1, whose only 'factors' are itself and 1. Factors, in this sense, are numbers that can be multiplied together to produce the number. Conversely, dividing a prime by any number other than itself or 1 will always produce a 'messy' result that is not a whole number. So 2 is prime since its only factors are 1 and 2 (1 x 2 = 2), 3 is prime because it divides properly only by 1 and 3 (3/2 = 1.5). The first few primes, therefore, are 2, 3, 5, 7, 11, 13, 17, 19, 23, 29, 31, 37...

2 The distribution of primes

Even at a glance, certain broad patterns among the primes are obvious. For instance, 2 is the only prime that is an even number (since the very definition of even numbers is that they divide by 2). Also, the frequency of primes decreases as numbers get larger, but maybe not as rapidly as you'd expect; there are 15 prime numbers below 50, 25 below 100, a total of 46 below 200, and some 168 in all the numbers below 1000. Mathematicians have been fascinated by patterns in the distribution of primes since the earliest times, as they offer a possible shortcut to finding them while reducing laborious calculations.

3 The fundamental theorem of arithmetic

Euclid's *Elements* (*c.* 300 BCE) contains some important ideas about primes. The first of these, today known as the 'fundamental theorem of arithmetic', states that every whole number greater than 1 is either a prime, or a composite made by multiplying a unique string of prime numbers (known as its 'prime factors') together. Based on this discovery, Euclid went on to show that there must be an infinite number of primes, based on the observation that if you take any particular list of primes, multiply them together and add 1, you will produce a number that is either itself an unlisted prime, or a composite whose list of prime factors has to include a prime not on the list. This means that however long the list gets, you can always prove the existence of one more prime beyond it.

4 Prime number sieves

One of the oldest and most popular methods of identifying primes is attributed to ancient Greek mathematician Eratosthenes. This 'sieve' is an algorithm or set of repeating instructions for finding all the primes below a certain number. The first step is to list all the numbers being investigated beginning at 2, the second to strike out every *n*th number on the list starting from *n*, where *n* is the lowest number not already struck out. Repeat or 'iterate' this process for the next number left unmarked, until you eventually reach an iteration on which no further numbers are struck out. For example, the first iteration strikes out all the even numbers above 2, the second all the odd multiples of 3, the third all the odd multiples of 5 that are not also divisible by 3, and so on.

	2	3	~~4~~	5	~~6~~	7
~~8~~	~~9~~	~~10~~	11	~~12~~	13	~~14~~
~~15~~	~~16~~	17	~~18~~	19	~~20~~	~~21~~
~~22~~	23	~~24~~	~~25~~	~~26~~	~~27~~	~~28~~
29	~~30~~	31	~~32~~	~~33~~	~~34~~	~~35~~
~~36~~	37	~~38~~	~~39~~	~~40~~	41	~~42~~
43	~~44~~	~~45~~	~~46~~	47	~~48~~	~~49~~

5 Hilbert's eighth problem

In 1900, David Hilbert included proofs of several ideas about primes as the 'eighth problem' on his list of unsolved mathematical mysteries. Chief among these are statements known as the Riemann hypothesis and Goldbach's conjecture. The Riemann hypothesis is the somewhat opaque statement that 'the Riemann zeta function delivers zeros only at negative even integers and complex numbers with real part ½. Goldbach's conjecture is much more straightforward; it states simply that every even whole number above 2 can be written as the sum of two primes.

6 The Riemann zeta function

So what is the 'zeta function'? In maths, a function defines a relationship between one or more objects that act as inputs (typically, but not always, numbers) and an output. The output of the zeta function is the sum of a mathematical series (a list of numbers) created by writing down all the numbers $1/n^s$, where n goes from 1 to infinity, and s is the input of the function. In other words, if n is 2 then the function is the sum of $1/1^2 + 1/2^2 + 1/3^2 + 1/4^2 + ...$ ($= 1 + \frac{1}{4} + \frac{1}{9} + \frac{1}{16} ...$). What makes the zeta function more complicated is that s can be a 'complex' number, a combination of 'real' natural numbers with 'imaginary' numbers based on i, the square root of -1 (a number that is incalculable in real terms but hugely useful in mathematics).

7 Riemann's hypothesis

The output of the zeta function is always 0 if s is a negative even integer (results called 'trivial zeros'). But Riemann's hypothesis says that it also delivers 'non-trivial zeros' if, and only if, s is a complex number of the form $\frac{1}{2} + ix$ (where i is the square root of -1). The hypothesis was proposed by German mathematician Bernhard Riemann in 1859, and the reason for its relation to prime numbers is that the zeta function appears in the 'prime number theorem', a formula that promises to calculate the number of primes below a specific number. In this formula, the real part of the zeta function at non-trivial zeroes controls the amount by which the location of primes varies around the locations that might be predicted from their general tailing-off in frequency at higher numbers.

8 Proving Riemann

In the century and a half since it was first proposed, attempts at proving the Riemann hypothesis have met with mixed results. Numerous techniques have shown that it is true for all complex values of s that match certain specific characteristics. What's more, thanks to the power of modern computers, it's possible to use algorithms that check the value of s for large numbers of non-trivial zeros. The most recent attempt to do this, in 2004, checked the first 10 *trillion* non-trivial zeros, as well as performing random 'spot checks' at levels up to around where the 10 *trillion trillionth* zero should be found. So far, the Riemann hypothesis has withstood every test, but from a mathematical point of view, that's not the same thing as a comprehensive proof.

9 Goldbach's conjectures

The easier-to-grasp second major element of Hilbert's eighth problem is usually taken as a statement that every whole number greater than 2 can be expressed as the sum of two primes, its so-called 'Goldbach partition' (that is, $n = x + y$ where n is greater than 2 and x and y are prime). German mathematician Christian Goldbach first proposed his conjecture (in a slightly different form) in a letter of 1742. Just to complicate matters, this form is also known as the 'strong Goldbach conjecture', distinguishing it from a 'weak' conjecture that every *odd* number greater than 5 can be expressed as the sum of *three* primes. The weak conjecture is so-called because it would automatically be proved if we proved the strong conjecture (if you know that any even number from 4 upwards is the sum of two primes, then you just have to add the prime 3 to it).

10 Solutions to Goldbach

Since it's more constrained in its scope, mathematicians have often focused on proving the weak conjecture. Early attempts at a proof relied on assuming the truth of the Riemann hypothesis, or could only be made to work for 'sufficiently large' numbers (a euphemism for numbers with at least 6.8 *million* digits, making checking all the rest by hand somewhat unfeasible!). A 1997 breakthrough showed that the Riemann hypothesis would prove the weak conjecture for all numbers with more than 20 digits (a low enough number for the exceptions to be checked with computers), and in 2013, Peruvian mathematician Harald Helfgott published a proof that did not rely on Riemann. Proofs of the strong version, however, remain elusive.

TALK LIKE A GENIUS

❛ There's one very good practical reason why all of this matters, and that's because a lot of internet security around the world is based on the idea of prime factors. Essentially, you create a very large number by multiplying two primes together, and then you can make that publicly available for sending encrypted messages to your website, but only the computer that knows the primes involved can decrypt the message. Anyone who grabs the data in transit might know the long number, but factorization (working out the two primes that were multiplied to get it) is the sort of task that would take centuries of computing time, so the system's secure. Any proof of Riemann might also point to a way to a faster factorization method, in which case we might be waving goodbye to e-commerce. ❜

❛ Prime numbers occur often in nature – one intriguing example is the periodic cicada. These insects spend most of their lives as grubs living in burrows, and then emerge to breed after either 13 or 17 years. There are two theories as to why they have this prime-numbered breeding cycle – it may be to make it difficult for predators to adapt their own breeding cycles to take advantage of such an abundant food source, or it could be so that two separate populations rarely overlap and therefore avoid competing for resources or breeding together and messing up the cycle. ❜

WERE YOU A GENIUS?

1 FALSE – prime numbers can also be divided by themselves (the result equals 1).

2 FALSE –although there is a decline, its rate varies significantly; the Riemann hypothesis offers a possible way of understanding it.

3 FALSE – however a proof of the hypothesis might point the way to calculating the 'prime factors' of large numbers.

4 FALSE – what's described is actually the weak conjecture.

5 TRUE – the quest for a general proof remains elusive.

All the known prime numbers follow a distinctive distribution, but proof that this is universal remains elusive.

Infinity

'There is a concept which corrupts and upsets all others. I refer not to Evil, whose limited realm is that of ethics; I refer to the infinite.'

JORGE LUIS BORGES

Infinity is one of the most remarkable and puzzling concepts in all of human thinking, with implications that stretch across mathematics, science and philosophy. At first glance, the idea that numbers, spaces and concepts can extend or increase forever without reaching an end seems straightforward enough, but when Georg Cantor put it on a mathematical footing in the 19th century, he raised more questions about the nature of infinity than he provided answers.

The implications of infinity are enough to make your head spin, even before you realize that it comes in different flavours.

1 Infinity goes both ways – both numbers and physical quantities can be infinitely added to, or divided an infinite number of times.

TRUE / FALSE

2 Georg Cantor described fundamentally different types of infinity – those we can at least begin to count in a meaningful way, and those we cannot.

TRUE / FALSE

3 Bijection is a process that creates gaps in infinite number sequences in order to fit more numbers in.

TRUE / FALSE

4 Hilbert's Hotel is a mathematical argument that uncountable infinities do not exist.

TRUE / FALSE

5 Fractals are patterns created by infinite division of relatively simple geometric shapes.

TRUE / FALSE

TEN THINGS A GENIUS KNOWS

1 Origins of infinity

The earliest person we know to have considered the concept of the infinite did so in a spatial, rather than a mathematical sense. Pre-Socratic philosopher Anaximander argued in the sixth century BCE that the cosmos had originated as a boundless, chaotic space called the *apeiron,* capable of generating and recycling infinite worlds, while Anaxagoras, a century later, believed the matter of the Universe was capable of infinite division. Aristotle, writing in the fourth century BCE, seems to have been the first to introduce an important distinction between the potential infinities of numbers and imagination, and the physical infinities proposed by his predecessors. While embracing the former, he rejected the latter and, in so doing, laid for the foundations for the spatially limited models of our Universe that persisted for almost 2,000 years.

2 Zeno's paradoxes

Some of the earliest and most important considerations of mathematical infinity come from Zeno of Elea, another pre-Socratic philosopher who wrote around 490 BCE and outlined a number of paradoxes involving what we would now call 'infinite series' (most famously, Achilles covering ground to catch up with a tortoise, but also an argument that if you measure it in a brief enough instant, a flying arrow is actually motionless). He sought to highlight the problem that since each step of an infinitely divisible task must take a finite amount of time, such tasks could never be completed. Chinese philosophers seem to have dabbled in similar ideas around the same time.

3 Infinite series

An infinite series is essentially a mathematical expression that 'expands' into an endless number of separate elements called 'terms'. Many series are divergent, growing larger with each term to give an infinite result, but a more useful group is convergent, with each term bringing them successively closer to a 'limit' that produces a fixed finite result. One obvious series is the sum of $(1/n)$ where n takes on values from 1 to infinity (in other words, $1 + \frac{1}{2} + \frac{1}{3} + \frac{1}{4} + ...$). Perhaps surprisingly, this turns out to be a divergent series with an infinite sum. In contrast, the infinite expansion of $(1/2^n)$, that is, $\frac{1}{2} + \frac{1}{4} + \frac{1}{3} + ...$ *is* convergent, and its sum is simply 1. The concepts of infinite series and limits play a vital role in the mathematical field of calculus – although Newton and Leibniz are usually given credit for discovering this in the late 17th century, much of the groundwork was laid by English mathematician John Wallis in a work of 1655 that also introduced the famous symbol ∞ as a way of representing infinity.

4 Set theory

Most modern thinking about infinity is founded in set theory, a mathematical approach based on applying rules of logic to collections of mathematical objects. German mathematician Georg Cantor single-handedly established this influential theory in a paper of 1874. Cantor introduced the idea of countable infinite sets; a set may have an infinite number of objects within it, but if they can be arranged in some kind of ordered list, each element can be made to correspond with a natural ('counting') number (i.e. 1, 2, 3, 4...), a process called 'bijection'. For example, the set of all integers (including negative ones) is countable because its elements can be ordered in a list beginning 0, 1, –1, 2, –2...

5 Uncountable infinity

In contrast to his countable infinities, Cantor also proved the existence of 'uncountable' infinite sets, whose elements *cannot* be put into an ordered list. Consider the set of *all* real numbers, including those with a decimal point followed by an arbitrary number of digits. Cantor proved this was uncountable by contradiction. First assume that the set of all real numbers is countable and a bijection can be made by putting each number in order of magnitude and matching it to a natural number. Cantor then provided a simple method for constructing a number that does *not* appear on the list (despite it apparently containing all real numbers), a contradiction that proves the initial assumption of countability wrong. In Cantor's terminology, uncountably infinite sets have a higher cardinality (number of elements) than countable ones; they are in some sense 'bigger'.

6 Counting the rational numbers

Perhaps surprisingly, the set of all rational numbers (integers and fractions of the form x/y, such as 1/4, and 13/72) is countable. You might think it was impossible to construct an ordered list of all possible fractions, but in 1878, Cantor showed that it could be done using an intermediate stage called a 'pairing function'. The idea is to put all possible pairs of natural numbers (the top and bottom halves of each fraction) in a coherent order where they can be made to correspond to a single counting number; the first few pairs in the list are (0,0), (0,1), (1,0), (0,2), (1,1) (2,0), (0,3), etc.

7 Hilbert's Hotel

In 1924, David Hilbert came up with an analogy that he felt demonstrated some of the peculiarities of infinite sets. He imagined a hotel with a (countably) infinite number of rooms, each of them occupied by a single guest. If a new guest arrives and asks to check in, how can they be accommodated? The answer, Hilbert argued, is to ask each guest to move into the next room along; the guest in Room 1 moves to Room 2, and so on. Because there is an infinity of rooms, no one finds themselves without a room, but Room 1 is now free for a new guest. A different moving scheme, asking everyone to move to a room whose number is double their initial one, can even allow the hotel to accommodate *infinitely* many new guests while remaining countable. Hilbert went on to show that it's even possible to fit an infinite number of guests from an infinite number of newly arrived coaches into the apparently full hotel, generating a series of nested infinities; the only limitation he found was that the number of these infinite levels must itself be finite.

8 Cantor's paradise

The 'higher-level' cardinality displayed by the set of all real numbers is known as the 'cardinality of the continuum'; one mental shortcut to imagining it might be to think of countable cardinality as marking off ticks on an infinite tape measure, and the cardinality of the continuum as trying to also take account of the infinite subdivisions between each mark on the tape. However, Cantor showed that mathematically, there must be an *infinite* number of possible cardinalities (something Hilbert later called 'Cantor's paradise'). The argument is that, just as any set has subsets that include some but not all of its elements, so it has a 'power set' that is the collection of all its subsets. Cantor showed that the power set *always* has more members than the original set, giving it a higher cardinality. By considering power sets *of* power sets, he constructed the idea of an infinity of different types of infinity.

9 Fractals

Made famous by chaos theory, fractals are patterns generated by the process of infinite iteration (repetition); the same patterns repeat again and again on ever-smaller scales. The first such pattern was identified by German Karl Weierstrass in 1872, but it was Cantor again who brought them to popular notice after identifying a means of dividing up a line segment in a fractal manner. Fractals can be generated from graphs of equations, but also by infinite repetition of fairly simple instructions on different scales – one curious factor is that the edge defining a fractal's boundary itself has huge complexity (a property mathematicians call 'fractal dimension'), while the area contained within the fractal shape dwindles to zero when looked at in enough detail.

10 Infinity in nature

Aristotle's dividing line between the potential infinity of mathematics and actual infinity in the physical world had a long legacy, and the idea of infinity in nature has only been widely accepted since the 20th century. Solutions to the field equations describing spacetime in Einstein's theory of general relativity allow for the possibility of singularities (physical equivalents of dividing by zero, where properties become infinite in extent); the best known of these are black holes (collapsed stars and other objects of infinite density), but the Big Bang itself was probably also a singularity. And while until recently most astronomers have been content to imagine the cosmos it created as finite in extent, it's now looking increasingly likely that our Universe is itself infinite, with different cardinalities of its own (see page 201).

TALK LIKE A GENIUS

'Most of Zeno's paradoxes use the same trick – subdividing a process again and again into an infinite number of steps, and then claiming that means that it can never be completed. Of course, it's a bit daft and I doubt Zeno himself ever tried to outrace an arrow, but if you *do* want to address it on its own terms, then you just have to point out that provided the time taken for each smaller division of a task reduces sufficiently, then even an infinite number of terms converges on a finite sum. That's why Achilles always catches up with the tortoise eventually.'

'Here's one of the weird paradoxes of set theory: you can make a set of all the natural counting numbers, and say that has a cardinality |N|. But you can also take just the even numbers, and if you do a bijection so that each corresponds to one of the natural numbers (so 2 matches 1, 4 matches 2, 6 matches 3, and so on), you end up showing that the set of all even numbers has exactly the same cardinality as the set of all numbers – despite their being half as many of them!'

'Amazingly, there's a fourth-century BCE text from the Indian Jain religion that talks about grouping numbers into three different broad classes and recognizes different levels of infinity and countability; Cantor was beaten to it by 2,200 years.'

WERE YOU A GENIUS?

1 FALSE – while infinity does indeed go both ways for numbers, many physical properties have a least-possible amount (a quantum) below which they cannot be divided.

2 TRUE – although Cantor also distinguished between other types.

3 FALSE – bijection is actually a process of 'matching' the natural (counting) numbers to the members of an infinite sequence in order to show that it is countable.

4 FALSE – the hotel analogy does not address uncountability.

5 TRUE – continued division in fractals creates complex, repeating patterns on many levels.

There are many different levels of infinity, some of which we can count, and some we can't.

Probability and statistics

'To understand God's thoughts we must study statistics, for these are the measure of His purpose.'

FLORENCE NIGHTINGALE

The abilities to correctly assess data (statistics), and to understand the likelihood or risk of particular events (probability), are vital skills for understanding an increasingly complex world. Yet these allied fields are poorly understood, and we are prone to reject them when they clash with our innate instincts. Any genius needs to have some grounding in the different ways that data is used, if only to spot when they're being misled by others, or fooling themselves.

Probability and statistics are mathematical tools for discovering truths about the real world – but can we handle the truth?

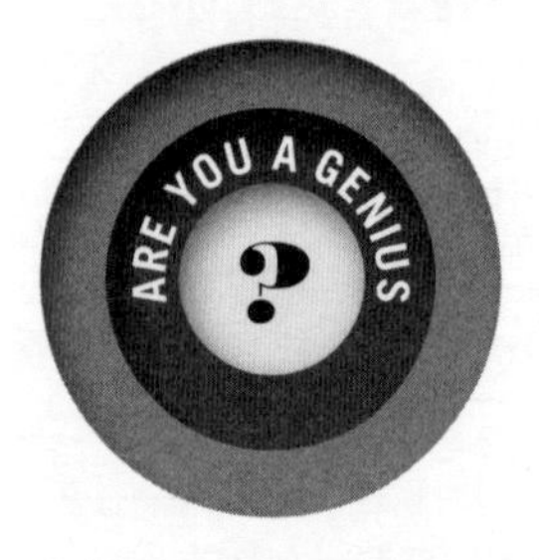

1 When rolling two six-sided dice, the probability of getting any specific combination of dots is 1 in 36.

TRUE / FALSE

2 When rolling the same two dice, the probability of getting any specific total of dots is 1 in 12.

TRUE / FALSE

3 Doubling the size of a statistical sample always lets you halve the margin of error.

TRUE / FALSE

4 Provided you bear in mind the larger margin of error, you can often get useful data from subsamples – specific sections with a survey or poll.

TRUE / FALSE

5 Average figures are a good way of representing statistical data.

TRUE / FALSE

TEN THINGS A GENIUS KNOWS

1 Probability

Most of us have an intuitive idea of probability; it's the likelihood of a specific event or outcome. Roll a dice, and the chances it will land on two are 1 in 6 – or 0.166, since mathematicians assign probabilities on a scale between 0 (impossible) and 1 (certain). Flip a coin, and the probability of it landing heads-up is 0.5. But as soon as problems become more complicated, most of us start to make mistakes; we assume links that aren't justified, and miss those that are. Flip an unbiased coin 85 times and it's *very* unlikely to land heads-up each time (the odds are 0.5 times 0.5, 85 times over). But those odds are just the same as for any other random 85-flip sequence. And even if the coin does land on heads each time, it doesn't have a memory; the next flip is still a 50/50 chance.

2 Surprising outcomes

An accurate understanding of probability can lead to some surprising revelations. A famous example is the 'birthday problem' – how many people do you need in a room before there is a 0.5 chance of two having the same birthday? The easiest way to think about this is to consider the inverse case where nobody shares a birthday – a probability that rapidly falls off as the number of people increases. Since the combined probability of all outcomes is always 1, a shared birthday becomes more likely than not when the probability of unique birthdays drops below 0.5, something that occurs for just 23 people.

3 Bayesian probability

Most forms of probability make calculations based on the information initially available about a given system. Eighteenth century English mathematician Thomas Bayes, however, developed an approach that modifies that knowledge using additional information gained along the way. At its heart is a probability law called Bayes' theorem, which allows new understanding to be factored into a prior model. Bayes' theorem offers one way of solving the famous 'Monty Hall' problem; if you're asked to select which of three boxes contains a prize, and are then shown that one of the unselected boxes is empty and offered the chance to switch your choice, should you do so? The answer is an emphatic yes: you will boost your chances of winning from 0.333 to 0.666. Since there was only a 1 in 3 chance that you picked the right box first time, being shown an empty box increases the odds of the other one holding the prize to 2 in 3.

4 Probability, uncertainty and risk

While properly assessed probabilities *should* underlie our perception of the likelihood and frequency of events, in reality we all too often make mistakes due to cognitive biases. Even setting aside the examples given above, we frequently overestimate the probability of rare but widely publicized events (both good ones like lottery wins and bad ones such as terrorist attacks) simply because we recall them better. We ignore background information such as the 'base rate' possibility of certain events in favour of new information that seems to override it, and we neglect the different time frames associated with different risks. For example, while you're extremely unlikely to die from being hit by an asteroid *in your lifetime*, some have estimated your *overall* probability of dying in such an impact as about the same as being killed by a tornado. This is because, while major asteroid impacts on Earth are incredibly rare, they have the potential to cause casualties on a truly vast scale.

5 Why statistics matter

Statistics is the business of data analysis – an array of mathematical techniques that can take numbers from limited surveys of the real world, identify patterns and trends within them, and assign probabilities that those patterns are broadly representative. Simple 'descriptive' statistics describes the basic features within a single 'data set' – for instance, the average and most common age of buyers for a particular product, and the spread of ages around those peaks. Inferential statistics, meanwhile, assesses possible relationships *between* data sets; does age affect the likelihood that someone is interested in the product? Statistics are a powerful tool both for finding trends in public opinion and for analysing scientific hypotheses, but it's important to remember that its conclusions are matters of probability, rather than certainty.

6 Sampling

In many statistical situations, access to data is limited by practical reasons; we cannot survey an entire population, map an entire area or repeat an experiment an infinite number of times. The challenge, then, is to create a statistically valid 'sample' that can reasonably be expected to mirror the characteristics of the overall subject under study. In opinion polling, for example, a sample is typically constructed with reference to socio-economic grouping, age and other patterns in the overall population known from comprehensive census data. In theory, it's possible to construct large samples that properly represent all these aspects for every part of the population, but in reality that can be a challenge.

7 Margin of error

A basic principle of statistics is to 'always show the error bars'. The techniques for calculating error are quite complicated, but the key point is that the bigger and better-constructed a sample or experiment, the more likely it is to accurately represent the underlying pattern. In polling, accuracy is expressed through 'margin of error' (MoE, a variation in the measurement either way). If a sample is truly representative, then its results should lie within the MoE of reality the vast majority of the time. However, even the best-constructed survey can occasionally produce unrepresentative results known as outliers, and statisticians express the likelihood of such results in terms of a percentage known as the confidence interval. So, for example, a properly constructed political opinion poll with a sample size of 1,000 should predict voting intention with a margin of error of 3 per cent and a 95 per cent confidence interval. That means there's just a 5 per cent chance of the true intentions of the population differing by more than 3 per cent from the survey results

8 Statistics, probability and proof

Researchers in science and many other fields often want to test the truth of a hypothesis through experimental results or statistical evidence, and probability plays a key role here. The standard technique is, in fact, to test the 'null hypothesis', a default assumption that there is no link between two data sets (or between one set and a predictive model describing a certain hypothesis). The resulting 'p value' measures the probability of the null hypothesis being true within a certain confidence interval. If the p value is below a certain threshold, called the significance level, then the null hypothesis can be rejected. This is not to say, however, that the proposed link or alternative hypothesis is necessarily proven true – further testing and experimentation are usually required to narrow down the possibilities.

9 Anecdote vs data

Just as with our assessment of risks, we frequently have to fight against our cognitive biases in order to properly assess statistical information. Our brains are naturally inclined to privilege personal experience and direct accounts over impersonal numbers, so that when a conflict arises, we usually doubt the statistical evidence. A common response to this states 'the plural of anecdote is not data' (and one only needs to look at the principles of MoE outlined above, to realise how unrepresentative tiny samples can become). But a more nuanced and persuasive view is that 'statistics are people, too'; in essence, surveys are indeed collections of anecdotes, filtered through systems that remove much of their detail and hopefully weight them correctly to represent reality.

10 Misuse of statistics

It's also worth remembering that statistical surveys are prone to misinterpretation and even outright abuse. Presenting participants with questions that carry an implicit interpretation of the 'facts' is a well-known tactic for pushing them to give certain answers, and we also have a natural bias towards giving 'positive' answers, so the way a question is framed can have a significant impact on the answer given. Other questions may not mislead participants, but can mislead our efforts to interpret the results – for instance, 'Have you or someone you know ever...?' questions are a common way of magnifying the strength of response and making issues look more significant than they are. Unfortunately, surveys are rarely an entirely disinterested academic exercise, and it's often the case that the commercial or political interests of a survey's commissioner shape its outcome (or at least the way it is reported).

TALK LIKE A GENIUS

‘Margin of error shows the real difference between anecdote and data. The MoE, even for a properly sampled large-scale poll, is usually around 3 per cent either way, but for 50 people it rockets to 14 per cent. That’s why you’re likely to end up frustrated if you think what your friends tell you is representative of widespread opinion and experience. But we still privilege anecdotes, of course, because we’re human.’

‘The base rate fallacy is a good example of how we get probabilities wrong. Suppose that black people make up 5 per cent of the population in a particular town, and there’s a disease there whose rate in the black population is *ten times* higher than in the rest of the community. Patricia is from the town and has the disease, so what are the chances that she’s black? The answer works out at just under 1 in 8 – if you thought it would be higher, that’s because you let the specific information about the disease override the ‘base rate’ information about how many black people there are to start with.’

‘US engineer Robert Howard came up with a clever way of showing the real risks involved in various activities in 1980, with a unit he called the ‘micromort’: a 1 in a million chance of death. Oxford statistician David Spiegelhalter added to that more recently with the ‘microlife’, a 30-minute reduction in life expectancy due to chronic risks such as drinking and smoking.’

WERE YOU A GENIUS?

1 TRUE – provided you keep a distinction between the two dice.

2 FALSE – the probability of each possible total depends on how many ways there are to reach it.

3 FALSE – increasing sample size has a big effect on MoE with small samples, but a much smaller effect when sample numbers get larger.

4 FALSE – this is only true if the subsample is also properly representative of its part of the overall population.

5 FALSE – averages are often used to skew perceptions because they disguise the distribution of data around them.

Probability is the surprising maths of likelihood and prediction, while statistics concerns how accurately limited data represents the real world, and how we can best present it.

Chaos

'The book of nature is written in the language of mathematics; without its help it is impossible to comprehend a single word of it.'

GALILEO

Chaos is one of those deceptive terms with a broad meaning in everyday conversation, but a very specific one for mathematicians and scientists. Understanding the difference between the two is important, because the mathematical form of chaos explains many of the limits of mathematical prediction – it's not that our models of phenomena such as weather fail because of inherent faults, but rather that we don't have (and perhaps can never have) a good enough picture of the conditions in which they start out.

In nature and in mathematics, it's becoming clear that chaos rules – can you become its master?

1 Chaos theory places limits on the accuracy of some mathematical predictions.

TRUE / FALSE

2 Any mathematical equation can give rise to chaos – it all depends on what natural phenomenon it is representing.

TRUE / FALSE

3 Chaotic attractors pull mathematical equations towards certain conditions, where they oscillate before eventually zeroing in on a final value.

TRUE / FALSE

4 Chaos theory is a direct contradiction of Newton's equations of motion in classical physics.

TRUE / FALSE

5 Astronomers think that chaos has had a major effect on the history of our solar system.

TRUE / FALSE

TEN THINGS A GENIUS KNOWS

1 Newton's predictable Universe

In 1687, Isaac Newton changed the way we see the world forever with the publication of his *Mathematical Principles of Natural Philosophy* (best known as the *Principia* and not to be confused with Russell and Whitehead's later work). In it, he set out laws of motion and gravitation that predict how objects act under the influence of different forces. Newton's laws form the basis of 'classical mechanics', and are still generally used today (although Einstein's relativity applies better in some extreme situations). Their discovery triggered a philosophical earthquake as they seemed to make the Universe deterministic; if you knew the initial locations and motions of all its particles, you could in theory predict its future.

2 The three-body problem

However, scientists soon discovered that a problem lies in wait for anyone trying to do this in reality. Take, for instance, the orbits of planets around the Sun; it's perfectly feasible to use Newton's laws to predict the motion of a single planet, but if you add another one (a third body in the system), the predictability rapidly fails. The changing balance of gravity between three large bodies causes each planet's orbit to show unpredictable variations when tracked for a sufficiently long period. Astronomers trying to predict the existence of hypothetical new planets in the 19th century struggled long and hard with this 'three body problem'. In 1880, French mathematician Henri Poincaré proposed that certain orbital configurations are simply inherently unpredictable, or what we would now call 'chaotic'.

3 The discovery of chaos

Despite doubts raised by the three-body problem, traditional 'linear theory' prevailed in ideas about physical systems until the middle of the 20th century. According to this model, small changes or unknown factors would produce equally small changes in results, rather than major variations. When results did not match those expected, the differences tended to be dismissed as measurement error or random variation ('noise'). A few mathematicians had an early inkling that something different might be happening, but it was only the development of computers, with their ability to run complex mathematical models many times over, that began to reveal the truth: many physical systems are far more sensitive to the precise initial conditions than anyone had expected.

4 The butterfly effect

In 1972, American mathematician and meteorologist Edward Lorenz published a paper on non-linear equations that introduced the most famous concept in chaos theory – the idea that a butterfly flapping its wings in Brazil can set off a tornado in Texas. Lorenz had first become aware of how sensitive some mathematical models can be in 1961, while running a primitive computer simulation of the weather. After laboriously keying in a starting condition as 0.506127 on the first run, the second time he merely entered 0.506, and was then astonished to discover that the simulation's results diverged wildly from the first run. This extreme 'sensitivity to initial conditions' is a hallmark of mathematical chaos, and it turns out to be surprisingly widespread.

5 Differential equations

Chaos is a feature that emerges from mathematical descriptions of change, known as 'differential equations'. Such equations usually describe the rate of change of one property in a system as another one changes – generically written as either dy/dx (the amount of change in y for a single-unit change in x) or simply $f(y)$ (with f as shorthand for a mathematical function). Linear differential equations calculate the rate of change in terms of simple additions, subtractions and multiplications of the variables involved, while non-linear differential equations involve raising variables to higher powers (for instance, time squared, or t^2). A key characteristic of chaotic systems is that they are always described by non-linear equations; raising the variables to higher powers magnifies the effect of small changes.

6 Chaotic attractors

One intriguing aspect of chaotic systems is that rather than being *entirely* unpredictable (with the potential to end up anywhere), many are unpredictable *within* fairly well-defined bounds, oscillating around one or more central states known as 'attractors'. This behaviour first emerged in Poincaré's work on the three-body problem, and in the case of planetary orbits, the attractor is easy to visualize as the median elliptical path around which the planet's actual position varies. With other systems, the attractor's attributes can be plotted on a two- or three-dimensional graph (for instance, the balance between predator and prey numbers in a particular ecosystem). Perhaps the most famous attractor is the butterfly-shaped pattern that emerges from Lorenz's atmospheric convection equation.

7 Weather prediction

The concept of chaos explains why predicting weather in even the medium term remains such a challenge. Earth's weather is a hugely complex system, driven by interactions of solar radiation, geography, ocean and atmosphere, and forecasts rely on models that combine the laws of thermodynamics with those of fluid dynamics. These equations are inherently chaotic; the way they develop is highly sensitive to initial conditions, and even with satellite monitoring and a widely distributed grid of ground stations, meteorologists can't hope to gather enough data for a single trustworthy prediction. Instead, the strategy is to run models a number of times, each with small variations in the starting conditions, and produce a forecast based on the most likely outcomes that emerge.

8 The Mandelbrot set

Fractals are intricately detailed patterns created by iterating mathematical algorithms at increasing levels of detail, and some have become icons of chaos theory because of the astonishing sensitivity to initial conditions they reveal. Most famous is the Mandelbrot set, discovered by Polish-born Franco-American mathematician and chaos theorist Benoit Mandelbrot in 1979 (he also coined the word 'fractal' itself). This famous chart marks out the set of complex numbers c that do not diverge towards infinity when a simple function $f(z) = z^2 + c$ is iterated with z increasing from 0 (the axes of the chart indicate the real and imaginary parts of c). Close to the boundary between diverging and non-diverging numbers, the tiny changes that make the difference between the two outcomes trace out incredibly complex and beautiful patterns. Often, chaotic attractors have such fractal structures.

9 The roughness of nature

Can chaos-driven mechanisms actually explain much of nature's apparent disorder, and render it not very chaotic, after all? Many scientists now believe this to be the case, and Mandelbrot's work was again hugely influential. He was particularly interested in the 'roughness' of nature, investigating apparently random phenomena, such as the shape of shorelines, the structure of plants and even the small-scale distribution of blood vessels. In all these, and more, he found patterns of self-similarity, repetition on different scales that mirrors patterns seen in fractals. By modelling the natural world, and even aspects of human behaviour, in terms of chaos-driven fractals, Mandelbrot seems to have discovered a hidden harmony in the natural world.

10 Self-organized criticality

An extraordinary offshoot of chaos is the concept of 'self-organized criticality' (SOC). First put forward by researchers in 1987, this is the idea that dynamical systems in nature inherently 'home in' on critical points at the boundary between two or more distinct 'phases' (significantly different arrangements of the system components). The critical point acts as an attractor, forcing the system to evolve towards it and oscillate around it, even if initial conditions were quite different. Proximity to such apparent equilibrium points is a common feature in complex systems ranging from financial markets and epidemics to evolution and solar activity, and SOC offers a potential explanation for how these delicately balanced states are reached.

TALK LIKE A GENIUS

❛ For all that it makes outcomes less predictable, chaos doesn't really undermine the fundamental principle of determinism – systems still evolve in line with the predictable laws of physics and biology, and cause follows effect; it's just that we don't have a sufficiently accurate model of the causes to work out the effects. I think Edward Lorenz put it best when he summarized chaos as the idea that "the present determines the future, but the approximate present does not approximately determine the future." ❜

❛ The butterfly effect's become something of a go-to storytelling device for writers, but it had a lot of precursors. Philosopher Johann Gottlieb Fichte wrote about the idea that moving even a single grain of sand would have unpredictable consequences affecting the whole beach. And I don't know if Lorenz ever read Ray Bradbury's *A Sound of Thunder* – the story where time travellers visiting the Late Cretaceous tread on a butterfly and come back home to find they're now being ruled by a fascist. ❜

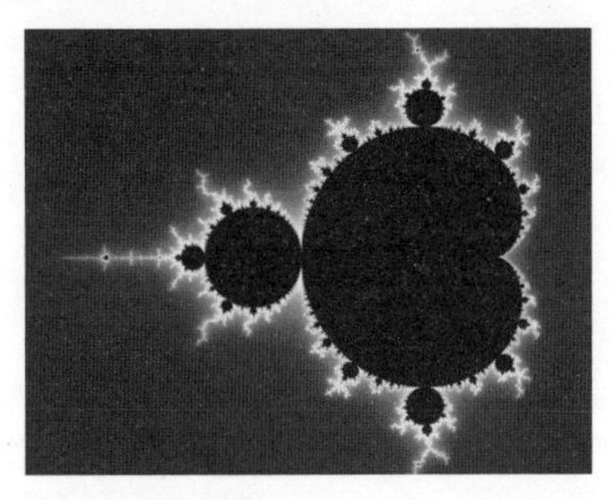

The Mandelbrot set

WERE YOU A GENIUS?

1 TRUE – chaos says that the accuracy of predictive models is dependent on the detail of the information we feed into them.

2 FALSE – chaos only arises in differential equations that include squares and higher powers.

3 FALSE – attractors never settle on a single final value, but continue to oscillate.

4 FALSE – chaos is an aspect of Newtonian physics rather than a contradiction of it.

5 TRUE – chaotic interactions between planets may have shifted their orbits in the distant past, and even brought them close to collision.

Chaos arises when slight differences in initial conditions give rise to very different outcomes – and it's more widespread than we thought.

Nanotechnology

'The principles of physics, as far as I can see, do not speak against the possibility of maneuvering things atom by atom.'

RICHARD FEYNMAN

Nanotechnology is the extraordinary science of the very small – construction on scales that range from nanometres (billionths of a metre), right down to custom assemblies of individual atoms a hundred times smaller. At present, nanotech is simply a specialized form of normal engineering and materials science, but according to some of its advocates, its ultimate potential could lie in the creation of tiny self-replicating robots, each one capable of carrying out a huge range of tasks.

Nanotechnology is already a part of all our lives, but is it really the transformative science that its promoters suggest?

1 The nanoparticles called 'fullerenes' are named after an American architect whose faceted domes have similar structures.

TRUE / FALSE

2 Nanoparticles are about 100 times thinner than a human hair, and can only be seen through powerful microscopes.

TRUE / FALSE

3 'Moore's Law' states that the density of components on a microchip currently doubles every two years. It's likely to continue at this rate for the foreseeable future.

TRUE / FALSE

4 Geckos use a natural form of nanotechnology to climb up smooth vertical surfaces. Engineers have borrowed this principle to create super-strong artificial. 'gecko tape'.

TRUE / FALSE

5 The accelerometers used inside smartphones and for triggering air bags during car crashes are one example of everyday nanotechnology at work.

TRUE / FALSE

TEN THINGS A GENIUS KNOWS

1 Room at the bottom

Nanotechnology is one of those cases where a wild prediction by a genius scientist eventually becomes a fully fledged reality. In December 1959, Richard Feynman (already renowned for his work on the theory of quantum electrodynamics) gave a talk at the California Institute of Technology, called *There's Plenty of Room at the Bottom*. In it, he set out the huge possibilities for manufacturing on the scale of atoms, even going so far as to offer thousand-dollar prizes for two feats of miniaturization (the reproduction of a printed page at 1/25,000 scale and construction of a working electric motor smaller than a 6 mm cube ($\frac{1}{64}$ in^3). The motor prize was claimed within a year, but Feynman had to wait until the 1980s for technology to catch up with his printing challenge.

2 Microelectronics and Moore's Law

In the meantime, the world underwent a revolution in microelectronics. Transistors (the gateways that control the flow of electric current in electronic devices such as computers), previously reliant on old-fashioned valves, were already being replaced with compact blocks of semiconducting silicon. In 1958, engineers at Texas Instruments figured out how to produce an entire circuit with multiple components and connections on a single silicon waver. As manufacturing techniques rapidly improved, a pattern became clear: Gordon Moore (later the cofounder of Intel) showed that the density of electronic components that could be fitted onto a single waver was doubling every year as the elements reduced rapidly in size. Moore's Law continues to hold to this day (even if the pace of advance has roughly halved since the millennium), and individual silicon elements are now being produced at a scale of just a few nanometres (billionths of a metre).

3 Top-down manufacturing

The techniques used for manufacturing silicon chips are so-called 'top-down' nanotechnology. They rely on creating components made out of relatively large numbers of atoms and making use of bulk properties, such as electrical conduction (which only really works when there are large numbers of electrons to push around in a material). Traditional chip manufacture uses 'photolithography', a technique in which light shone through a photographic mask onto light-sensitive chemicals 'etches' components onto an underlying substrate (usually silicon). Circuit elements are not usually required to move, but the same technique can also be used to create miniature mechanical components, such as wheels and drive shafts.

4 Bottom-up nanotech

Even more impressive applications may await at the other end of the nanotechnology scale. The 'bottom-up' approach involves manipulating individual atoms on a substrate to store data or build molecular machines. Such manipulation only became possible in the 1980s, after researchers at IBM in Zurich discovered the principle of the scanning tunnelling microscope (STM). A step beyond the ordinary electron microscope, the STM harnesses an effect called 'quantum tunnelling' that can not only offer a direct view of individual atoms on a surface, but also allows them to be picked up on the microscope's needle-like probe, and redistributed.

5 Carbon nanostructures

One example of nanotechnology already being put to work are the molecules known as 'fullerenes'. These balls and tubes contain hundreds of carbon atoms in various different geometric configurations. Fullerenes form naturally in the atmospheres of

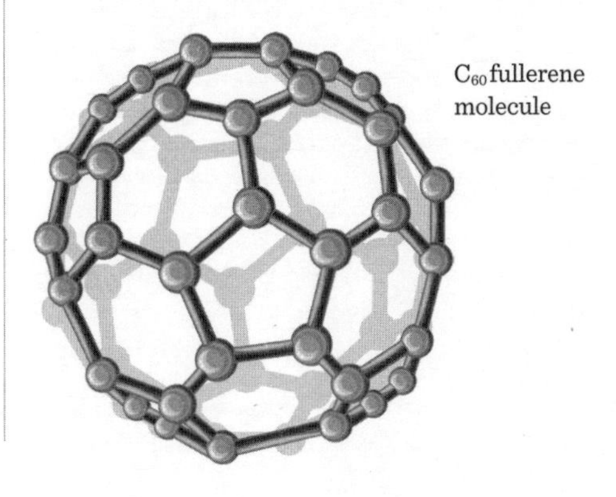

C_{60} fullerene molecule

dying stars, but they can also be made in a laboratory by evaporating graphite with a high-powered laser beam. In bulk, their lattice-like structures can be 20 times stronger than steel, but half the weight of aluminium, and with manufacturing costs rapidly falling they're likely to be the engineering materials of the future. Other fullerene properties, such as chemical inactivity and the ability to conduct both heat and electricity with extreme efficiency, are also being harnessed; for example, fullerenes can be used as 'containers' to deliver toxic drugs to specific parts of the body in certain cancer therapies.

6 Nanoengineering

In 1986, American engineer K. Eric Drexler wrote a bestselling book, *Engines of Creation*, that kicked nanotechnology into the mainstream. The term 'nanotechnology' had been coined by Japan's Norio Taniguchi as early as 1974, but Drexler was the first person to highlight many of the speculations in Feynman's long-overlooked talk. Principally, he argued for the potential of tiny machines he called 'assemblers', nano-scale devices that would harvest individual atoms and molecules from the environment and process them in various ways. Potential applications include neutralizing pollutants by binding them into unreactive molecules, extracting useful metals from ores, and even building larger-scale components and machines.

7 Von Neumann machines

For many nanotech advocates, the ultimate goal is to create self-replicating nanomachines. Named 'von Neumann machines' after Hungarian-American polymath John von Neumann (1903–57), these devices would effectively be a form of artificial life, capable of finding materials to keep running and reproduce themselves from the environment. Doubts were raised, however, by Nobel-winning chemist Richard Smalley (one of the discoverers of fullerenes), who argued that assemblers would only be able to function practically in a liquid environment provided with the right 'nutrients'.

8 Small-scale physics

Nanotechnology isn't just about miniaturizing principles that work on larger scales, however, because individual atoms and small collections don't necessarily behave in the same way as bulk materials. For example, normally unreactive metals, such as platinum, can be turned into powerful chemical catalysts by using nano-scale grains to display a large surface area to the surrounding environment. Other materials change their optical properties (copper is transparent on nano-scales) or show different electrical properties (silicon, usually an insulator, is an electrical conductor on the tiniest scales). Nanoengineers have to take account of these unusual behaviours, but can also turn them to their advantage.

9 Nano-applications

Current uses of nanotechnology often involve coating other materials in fine particles; for instance, surfaces can be coated with nanoparticles that make them resistant to dirt and damage. Nano-sized titanium oxide grains in sunscreen, meanwhile, provide an efficient barrier to UV rays that works for much longer than conventional sunscreens. Other products include highly efficient filters made from a mesh of countless nanofibres, and more efficient catalytic converters for cars. More sophisticated nanomachines using the top-down approach are still on the drawing board, but they promise to perform far more complex tasks; one widely proposed application is in medicine, where nanorobots injected into your body could hunt down and eradicate diseased tissues, or even help repair damaged organs.

10 Dangers of nanotechnology

Some environmentalists and even sceptical scientists are concerned that nanotechnology could easily slip from our control to become a threat. On a slightly larger scale this is already happening, with growing evidence of the effects from tiny plastic microgranules from cosmetics and other products escaping into the oceans and concentrating in the marine food chain. The idea of self-replicating von Neumann machines is therefore even more of a worry – their numbers could grow exponentially in the presence of suitable 'food', and even if they were released with the best of intentions, recent history sadly provides plenty of examples of environmental disasters after apparently harmless new species were introduced into delicately balanced ecosystems.

TALK LIKE A GENIUS

‘ Feynman’s writing prize ended up being claimed by a Stanford grad student called Tom Newman. He etched the first page from Dickens’ *A Tale of Two Cities* onto a 200-micron square of plastic with an electron gun – at that scale you could fit the entire *Encyclopedia Britannica* onto a pinhead! ’

‘ Supporters argue that von Neumann machines would only have benign functions, and anyway you could fit them with some sort of “kill switch”. But what happens if the replication process is less than perfect and the controls don’t work? You could even imagine a situation in which the machines start to “evolve” beyond our control by accumulating and passing on random errors. Realistically, the obvious way to get around the problem is not to build a completely self-sufficient machine in the first place. Drexler was probably his own worst enemy when he first speculated about runaway replicators turning the entire planet to “grey goo”. ’

WERE YOU A GENIUS?

1 TRUE – Richard Buckminster Fuller invented so-called ‘geodesic’ domes. Spherical C_{60} molecules are also nicknamed ‘buckyballs’.

2 FALSE – a human hair is actually *c.* 100,000 nanometres wide, dwarfing true nanostructures. ‘Bottom-up’ nanostructures are invisible even to electron microscopes.

3 FALSE – there’s a lower limit to the size of a practical electronic circuit. Engineers think Moore’s Law will cease at about 1.5 nm scales.

4 TRUE – gecko toes are covered in nanoscale ‘spatulae’, whose large surface area boosts the power of normally weak intermolecular forces.

5 FALSE – current accelerometers, tiny though they may be, are ‘just’ microtechnology.

Harnessing the world of small-scale technology opens up new horizons for energy efficiency, environmental engineering and even robotics.

Quantum physics

'Anyone who is not shocked by quantum theory has not understood it.'

ATTRIBUTED TO NIELS BOHR

For a scientific theory that's been around in one form or another for more than a century, and which underlies much of modern technological society, it's remarkable how daunting quantum physics still appears to be to most people. In part, that undoubtedly owes much to the intimidating equations physicists use to put quantum theory to work – but if you strip those away, it's easy to boil the field down to a handful of key ideas. The real challenge lies in how these ideas reveal a world fundamentally at odds with our everyday experience.

Quantum physics reveals that the Universe is not just stranger than we imagine, but stranger than we *can* imagine...

1 While normally experienced as a wave, light is also split into tiny particle-like packets called photons.

TRUE / FALSE

2 Although usually treated as solid, point-like objects, subatomic particles can manifest wavelike behaviour.

TRUE / FALSE

3 Quantum physics allows us to work out the chemical composition of distant stars.

TRUE / FALSE

4 While quantum theory predicts that space is filled with virtual particles popping in and out of existence all the time, that idea is still hypothetical.

TRUE / FALSE

5 Physicists have detected wavelike behaviour in particles that are one-thousandth of a millimetre across.

TRUE / FALSE

TEN THINGS A GENIUS KNOWS

1 What is quantum theory?

Quantum theory describes how, on tiny subatomic scales, physics does not behave in the deterministic, cause-and-effect manner we would expect from classical models (the simple but elegant equations worked out by Isaac Newton in the late 17th century). The word 'quantum', in essence, means 'small bits', and quantum theory relies, at its heart, on the idea that certain phenomena, while appearing continuous on everyday scales, are actually broken down into tiny packets, or 'quanta'. This explains why a whole range of physical phenomena, from light to radioactive decay, do not behave in the way we might otherwise expect.

2 Trouble with light

Quantum physics arose from the same debate about the nature of light as did relativity (see page 193). In 1865, James Clerk Maxwell showed that visible light could be described as an electromagnetic wave, a model that predicted the existence of similar waves with different wavelengths, such as radio waves. Modelling light sources, however, raised a different problem: how did the range of different wavelengths emitted by an idealized light source (a so-called 'black body') vary with temperature? Physicists discovered equations to accurately describe behaviour both at low temperatures and long wavelengths and at high temperatures and short wavelengths – but they couldn't make the two work together, a problem later called the 'ultraviolet catastrophe'.

3 Birth of the photon

In 1900, German physicist Max Planck found an ingenious way to cheat his way around the ultraviolet catastrophe – if he assumed that a black body was producing light in small discrete packets whose energy was linked to their wavelength, then the patterns of energy output fell into place. Planck suggested this pattern of emission was somehow a result of behaviour in the light-emitting material, but in 1905 Albert Einstein went one step further, arguing that quantized packets (today known as 'photons') were fundamental to the nature of light itself. This could explain why electric currents run through certain metals when they are exposed to weak short-wavelength light, but not when they are illuminated by intense long-wavelength light: the so-called 'photoelectric effect' depends on the energy packaged within individual light quanta, rather than the number of quanta striking the surface.

4 Wave-particle duality

Photon theory suggests that light has both particle- and wavelike properties – an idea that resolved many questions about the behaviour of light. In the mid-1920s, physicists began to wonder if something similar might explain problems in the behaviour of atoms and subatomic particles. Frenchman Louis-Victor de Broglie suggested that particles might have their own 'de Broglie wavelength' associated with their momentum. The wavelengths of even the smallest particles would be so short (much shorter than the wavelengths of visible light) that they would normally appear as point-like objects, but in certain situations the wavelength could manifest itself and even be put to use. A classic demonstration is the way that electron beams spread out and create interference patterns after passing through narrow slits, but this is also the principle behind the electron microscope, which uses the ultrashort wavelengths of electrons to image objects that cannot be sharply resolved with light waves.

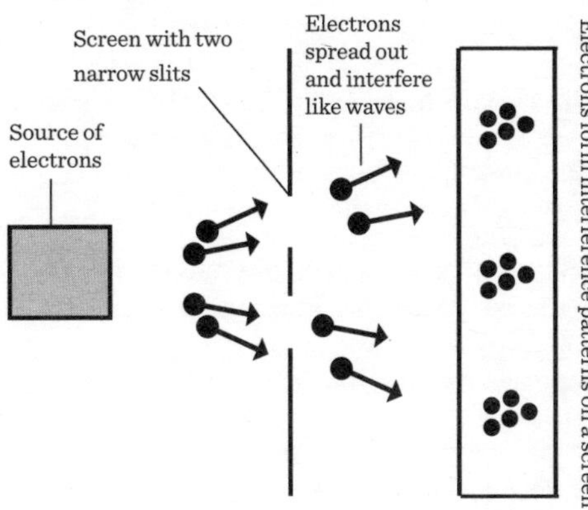

5 The wavefunction

An object's wavelike properties are described by its wavefunction (denoted by the Greek letter *psi*, Ψ) – an equation that basically describes the distribution of the particle's properties in space and time, first worked out by Erwin Schrödinger. You can plot the wavefunction as a graph showing how its strength varies with position – the simplest forms show a bell-like distribution with a single peak smoothly trailing off to zero. Physicists have some pretty fundamental disagreements about what the wavefunction means, but for practical purposes, it usually describes the probability of observing the object's properties in a certain state (for example, the position of a particle, or its energy) when a measurement is made. The spread-out nature of the wavefunction means that a quantum particle's properties are never entirely deterministic; there's always a chance that they will vary from the 'most likely' properties predicted by classical physics.

6 The quantum atom

One obvious area in which quantum effects make their presence felt is the structure of atoms. Traditional models explained atomic structure with an atomic nucleus of heavy particles (protons and neutrons) surrounded by low-mass electrons zipping around in shells called 'orbitals'. However, it turns out that a model in which the electrons are 'fuzzy', with their properties mostly smeared out around the orbital, does a much better job of describing how atoms behave in reality. That's important because the characteristics of electrons influence how atoms interact in chemical reactions.

7 Radioactive decay

Quantum effects are essential to describing radioactive 'alpha decay', in which clusters of protons and neutrons escape from an atomic nucleus, releasing energy and transforming the atom to a different element. Conventional physics says this should be impossible owing to an energy barrier around the nucleus, but quantum physics means there is always a possibility of the alpha particle's wavefunction reaching beyond the barrier; in other words, gaining enough energy to tunnel though the barrier and break free. It's impossible to predict just when an individual 'quantum tunnelling' event of this kind will happen, but with a substantial sample of material you can work out the time it takes for half of the atoms to decay – the so-called 'half-life'.

8 The uncertainty principle

The unpredictability of the quantum world is summed up in Werner Heisenberg's famous 'uncertainty principle', which states that it's impossible to know both values in certain pairs of 'complementary' variables with an arbitrarily high precision. In other words, the more tightly you pin down one variable, the less accurately you can estimate the other. One example is the position of a particle and its momentum (mass times velocity). Assuming the particle's mass is fixed, this means that the more accurately you measure the particle's location, the less accurately you can know its speed.

9 Vacuum energy

Another pair of complementary variables is the energy present in a region of space, and the time over which that energy is fixed. The so-called 'time-energy uncertainty relation' allows the Universe to create small amounts of energy from nowhere, provided they are put back almost instantaneously – the larger the amount of energy, the less time it can be borrowed for. That might sound like science fiction, but in fact it's a real phenomenon – empty space is full of short-lived 'virtual particles' popping briefly into existence using this borrowed energy, and in extreme situations (for instance around black holes), these particles can become persistent or 'real'.

10 Quantum theory in practice

Hopefully we've already explained, to some extent, why quantum theory is important to scientists. But it also lies at the heart of a whole range of modern technologies. Lasers, microelectronics and a variety of medical imaging techniques are all based on quantum principles. Nuclear power stations and even solar panels could not be built if it were not for our understanding of the way atoms and subatomic particles behave. And as we'll see in the next chapter, some of the stranger aspects of quantum theory have even more powerful applications.

TALK LIKE A GENIUS

‘That smartphone camera you’re so fond of is another example of quantum physics at work – it’s based on a semiconductor chip that counts the packets of light striking its individual pixels.’

‘The de Broglie wavelength of any object with mass is equivalent to a really tiny number called the “Planck constant”, divided by its momentum. That means that the wavelength is immeasurably small for anything much larger than an electron, and explains why everyday objects don’t normally show wavelike properties.’

‘You need to be careful not to confuse uncertainty principle with the measurement problem. The first is a fundamental limit on how accurately you can know stuff, while the second is the more practical problem that sometimes you can’t measure one aspect of a system without changing something else. In theory, you might be able to come up with ways of working around the measurement problem – but you can never get around the uncertainty principle!’

WERE YOU A GENIUS?

1 TRUE – individual light photons can be detected using devices such as electronic cameras.

2 TRUE – small particles can be made to display interference, uncertainty in position, and other wavelike behaviours.

3 TRUE – the atomic structure determined by quantum physics causes atoms to emit and absorb light at specific energies, creating a chemical fingerprint in the light of distant objects.

4 FALSE – virtual particles produce a measurable force known as the ‘Casimir effect’.

5 FALSE – so far, the largest objects in which wavelike properties have been measured are balls of carbon one nanometre across.

On the smallest scales, waves behave like particles and particles like waves. As a result, the Universe is a lot less predictable than we once thought.

Schrödinger's cat

'The task is, not so much to see what no one has yet seen; but to think what nobody has yet thought, about that which everybody sees.'

ERWIN SCHRÖDINGER

How does the inherently unpredictable subatomic world of quantum physics interact with the apparent predictability of the large-scale world? If every atom contains subatomic particles in a fuzzy state of uncertainty, described by interfering wavefunctions, then why does the real world obey boring certainties such as the rules of cause and effect, and the laws of Newtonian physics? The most famous thought experiment in scientific history highlights the paradox, even though it doesn't offer any solutions in itself.

There are several ways of reconciling quantum physics with the everyday world – but which one is right?

1 Quantum teleportation – the sending of data in the interlinked states of quantum particles – is a promising idea, but so far it's only been tested in laboratories.

TRUE / FALSE

2 Quantum computers are already capable of solving problems up to 10,000 times faster than conventional supercomputers.

TRUE / FALSE

3 Although they've never done it to anything as large as a cat, scientists have observed the effects of quantum uncertainty in objects that are large enough to see with the naked eye.

TRUE / FALSE

4 Quantum entanglement could break the law of cause and effect, by allowing information to be sent faster than light.

TRUE / FALSE

5 If the 'Many Worlds' interpretation is correct, every quantum measurement we make causes the Universe to split into different realities.

TRUE / FALSE

1 The wavefunction problem
The heart of the problem in uniting the quantum and everyday worlds lies in a phenomenon known as the 'collapse of the wavefunction'. Quantum systems, such as electrons and atomic nuclei, have wavelike as well as particle-like properties, with the wavelike bit best described by an equation called the 'wavefunction', which describes the probability of a system having a particular set of properties (for instance, position in space or momentum) at a particular time or location. We can describe the wavefunction mathematically, but whenever we observe a quantum system, we always find it in one particular state, with a distinct position, momentum and so on – it's never spread out. So what's going on?

2 The Copenhagen Interpretation
The first (and still the most popular) answer to that question, known as the 'Copenhagen Interpretation', was developed by Niels Bohr, Werner Heisenberg and others in Denmark in the early days of quantum physics. According to Copenhagen, the very act of measurement, and the interaction that implies with the larger-scale world of laboratory apparatus and scientists, causes the wavefunction suddenly to collapse onto a particular, specific outcome. Strictly speaking, Copenhagen says that it's meaningless to consider the system's particle-like properties until a measurement is made: the wavefunction describes the probability of certain properties appearing during measurement, but in theory even the least likely outcomes are possible.

3 Schrödinger's experiment
Erwin Schrödinger, father of the wavefunction itself, didn't think much of the Copenhagen Interpretation, despite its popularity. In order to highlight its absurdities, he imagined an experiment that would magnify quantum uncertainty to a macroscopic scale: placing a cat in a box with a vial of poison designed to break if a small radioactive source emits a particle in a certain time. Since the radioactive decay event is quantum in nature, Schrödinger argued, its wavefunction should not collapse until it is observed – in other words, the particle remains in a 'superposition' of decayed and undecayed states until it interacts with the outside world through measurement. By extension, this should then mean that the poison vial is both broken and unbroken, and the cat both dead and alive at the same time.

4 Solutions to Schrödinger
Schrödinger's experiment was purely hypothetical – it's needlessly cruel to the poor cat, and even if you did try to replicate the set-up you wouldn't learn anything because you can't see what's going on inside the box without opening it and therefore bringing the experiment to an end. But the problem it highlights is a real one – what stops everyday objects from falling into a state of quantum uncertainty? Supporters of Copenhagen get around the problem by arguing that the cat itself is an observer, but that's not the only solution, and a lot hangs on what you think the wavefunction actually is, as described below.

5 Is the wavefunction real?
Broadly speaking, there are two ways of interpreting the wavefunction. One is that it's purely a mathematical tool – a description that allows us to work out the odds of a quantum system showing particular properties when we measure it. According to this reading, quantum systems really have discrete, particle-like properties at all times – they just appear fuzzy because we can't be everywhere, measuring

everything at once. The other interpretation is that the wavefunction is a real thing – when we're not looking, a particle's properties really are smeared out across space and subject to quantum uncertainty. Hardcore Copenhagen advocates say the reality doesn't really matter one way or another – we should just 'shut up and calculate!'

6 The many worlds interpretation

The most famous alternative to the Copenhagen Interpretation builds on the assumption that not only is the wavefunction real, but it is the fundamental stuff of the Universe itself. According to the 'many worlds interpretation', formulated by Hugh Everett in the 1950s, any observation of a quantum system (direct or implicit) causes the Universe to branch off into multiple parallel realities – one for each of the possible outcomes. In other words, the act of opening the box creates two Universes, one for each of the cat's possible fates. This mind-boggling idea implies an infinite number of Universes existing side by side in a complex, many-dimensional 'multiverse', with every possible outcome happening somewhere.

7 Objective collapse

One problematic aspect of the Copenhagen Interpretation is that it brings subjectivity to the Universe – wavefunction collapse requires measurement, and measurement requires a conscious observer. This gives rise to some interesting cosmological questions, but a lot of scientists feel a bit uncomfortable around the implication that the Universe might dissolve into quantum uncertainty if we all stopped looking at it. 'Objective collapse' is an interpretation that avoids privileging observers – instead, it suggests that the wavefunction is inherently leaky or 'decoherent' – it naturally collapses as it interacts with the larger-scale Universe around it. In this interpretation, therefore, the cat's fate is resolved long before the box is opened.

8 Entangled particles

An extraordinary piece of evidence for the reality of the wavefunction and something like the Copenhagen Interpretation comes from the phenomenon of 'quantum entanglement'. Put simply, in certain conditions it's possible to create pairs of particles that we know must have complimentary quantum properties, without measuring the properties themselves. Such pairs are said to be 'entangled', and can be described by a single wavefunction that remains uncollapsed until one or the other particle is measured. The weird thing is that even if the particles are separated from one another, the unmeasured particle somehow 'knows' that its counterpart has been outed, and immediately takes on the complimentary property. Einstein called this freakish behaviour 'spooky action at a distance'.

9 Teleportation

The collapse of entangled particle pairs happens instantaneously and without any conventional 'communication' between particles. As such, it has huge potential as a technology for sending data, so-called 'quantum teleportation'. For example, you could create a pair of electrons that spin in opposite directions and send one off on a spaceship to a distant interplanetary colony – when the particle kept on Earth is measured, the other one will instantly collapse into the complimentary state, allowing the colonists to know the spin of its Earthly sibling. Of course, the spin of a single pre-prepared electron isn't exactly earth-shattering information on its own – the trick is to find ways of manipulating the paired particles to send data without permanently collapsing their wavefunctions.

10 Quantum computing

Entangled particles also have the potential to revolutionize computing. Where electronic computers rely on data 'bits' that take on values of '0' and '1' one calculation at a time, a quantum computer could store 'qubits' of data in the spin of electrons or the vibrations of light photons. And so long as their wavefunctions remain uncollapsed, these qubits could take on values of 0, 1 or anything in between *at the same time*. This property of superposition allows a quantum computer to work on many computations simultaneously, with the potential to solve mathematical problems millions of times faster than conventional supercomputers.

TALK LIKE A GENIUS

“Once quantum computers move from being laboratory playthings to real-world machines, you can kiss goodbye to conventional ideas about electronic privacy. A lot of internet encryption is based on shared keys – strings of numbers that are themselves the multiples of two primes. Conventional computers would take ages to work out those “prime factors” by brute force, but a quantum computer could find them in seconds, and a lot of experts think we’ll have quantum computers powerful enough to do that in the next decade.”

“Ironically, the best way of protecting against hacking by a quantum computer would be to send the information in a quantum form – if you’ve got a pair of entangled particles you could send information by conventional means in a code that only works if you know the quantum state of the sender’s particle. You “send” the key by measuring the particle – the entangled wave function collapses and the receiver can work out the state of the sender’s particle. It’s beautiful in principle – but a bit of a pain for the man in the street who just wants to send his credit card details securely!”

WERE YOU A GENIUS?

1 FALSE – teleportation has actually been done over distances of more than 500 km (300 miles).

2 TRUE – the D-Wave 2000Q, launched in 2017, has 2,000 qubits and can outpace algorithms used on conventional supercomputers.

3 TRUE – in 2010, researchers at the University of California created a tiny resonator, a tuning-fork-like device just 0.04 mm long, and successfully put it into a quantum superposition.

4 FALSE – because there’s no way of influencing the state into which each particle collapses, you can’t use entanglement to send information faster than light.

5 TRUE – however, according to the theory, the splits propagate out across the Universe at the speed of light.

Quantum uncertainty usually disappears in the everyday world, but preserving it can produce some remarkable effects.

The Higgs boson

'It is probably easier for believers to recognize the importance of what has taken place if it is called the "God particle". But the name itself is a sham, it was a joke, you know?'

PETER HIGGS

Often called the 'God particle' and theorized since the 1960s, the Higgs boson was the missing piece in the so-called 'Standard Model' of particle physics, helping to solve the otherwise tricky question of how particles get mass. In order to really understand what it is, though, we first need to get to grips with a whole array of confusingly named particles that can seem daunting to a non-genius. Fortunately, beneath all the strange names and mysterious properties lies a pattern of surprising simplicity and elegance.

The Higgs boson is the missing piece of a jigsaw that's only been found after a century of research. But do we know what the full picture is even now?

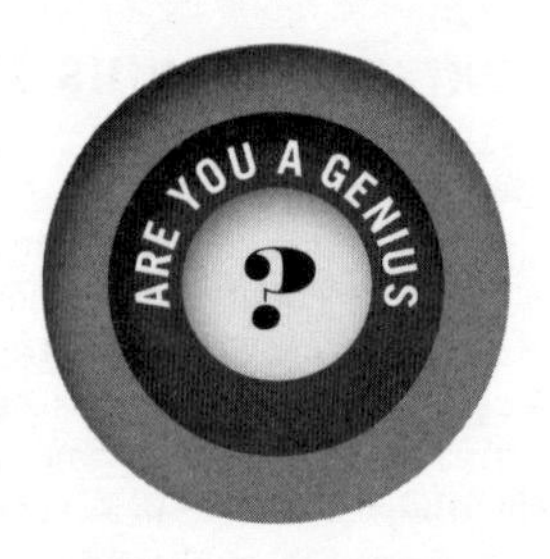

1 In 1905, Albert Einstein became the first person to identify visible evidence for the existence of atoms, even though it was impossible to 'see' atoms themselves until the 1980s.

TRUE / FALSE

2 Packed with positively charged protons, the only reason that atomic nuclei don't fall apart as their similar charges repel, is the presence of uncharged neutrons that allow the protons to keep a greater distance.

TRUE / FALSE

3 All atoms of a given element have a set number of protons and neutrons in their nucleus.

TRUE / FALSE

4 In our part of the Universe, antimatter is only naturally created by nuclear reactions, which is lucky, as it means there isn't any around to destroy normal matter.

TRUE / FALSE

5 Theories suggest there could actually be as many as five different Higgs particles, so four more are still waiting to be discovered.

TRUE / FALSE

TEN THINGS A GENIUS KNOWS

1 Elements and atoms

Elements are the basic building blocks of matter; substances that display unique chemical properties and that cannot be split apart into other substances by chemical reactions. Advances in chemistry have revealed that some materials once thought to be elements are in fact compounds, while new elements have been discovered to take their place. All elements are made up of individual atoms whose unique properties distinguish them from the atoms of other elements in both appearance and behaviour; their existence was speculated by ancient Greek philosophers, but only proven beyond doubt in the 18th century, when experiments showed how elements always react together in set proportions. Today, some 118 different elements are known, though only 90 of these occur naturally on Earth.

2 Inside the atom

Early last century, physicists established that all atoms are actually made up of three smaller particles – electrons, protons and neutrons. Electrons are lightweight but carry negative electric charge – ignoring quantum uncertainty, we can think of them as tiny satellites orbiting a compact central nucleus that carries the vast majority of the atom's mass. The nucleus is made up of high-mass protons with positive charge, and uncharged neutrons of similar mass. Each element is defined by its 'atomic number', the number of protons in its nuclei. In an electrically neutral atom, this is also the number of electrons orbiting around the nucleus. Electrons orbit in a complex series of shells and subshells, and atoms with full or half-full outer shells tend to be the most stable forms, and most chemical reactions involve either the exchange or sharing of electrons in the outer shell in order to achieve stability.

3 Atom smashing

Particle accelerators are the primary tool used by physicists studying the deep structure of matter. Machines such as the Large Hadron Collider (LHC) use powerful magnetic fields to accelerate electrically charged particles around a doughnut-shaped track until they are moving close to the speed of light itself. As the particles collide, they transform into pure energy in accordance with Einstein's famous equation $E=mc^2$. This energy rapidly condenses back into particles, and the extreme conditions thus created briefly allow particles to exist that cannot survive on their own in the lower-energy conditions of the present-day Universe.

4 Quarks

Atom-smashing experiments have shown that only one of the three main subatomic particles – the electron – is truly indivisible or 'elementary'. Both protons and neutrons can be broken up into even smaller particles called 'quarks' (named by discoverer Murray Gell-Mann from a nonsense phrase in James Joyce's *Finnegan's Wake*). There are six types or 'flavours' of quark in total (up and down, charm and strange, top and bottom, in order of increasing mass), but only the up and down quarks are found in everyday matter. Up quarks carry an electric charge of +2/3 (compared to an electron's –1), while down quarks carry a charge of –1/3, and each quark has a mass about 1/3 of a proton's (about 612 times that of an electron). Combining two up quarks and one down makes a proton, while two downs and an up make a neutron.

5 Leptons

Further experiments have shown that, by analogy with the quarks, the electron is just the 'everyday' form in a family of three similar particles (the others are called the 'muon' and the 'tau'), and each of these particles has a counterpart 'neutrino', a small uncharged particle with no charge and very low mass. Together, these pairs form three 'generations' of elementary particles called 'leptons', analogous to the three paired generations of quarks. Leptons and quarks together form the building blocks of all the 'normal' matter in the Universe.

6 Antimatter

As if this wasn't complicated enough, it turns out that each matter particle has an evil twin – a counterpart with equal, but opposite, electric charge and other quantum properties such as spin (see below). Antimatter crops up a lot in sci-fi (most famously as the power source of *Star Trek*'s warp drive), on account of its most famous behaviour: when matter and antimatter particles make contact with each other they disappear or 'annihilate' in a blast of pure energy that manifests itself as a burst of gamma rays. The fact that we're still here and aren't constantly being irradiated by deadly radiation is testament to the fact that antimatter isn't common in our part of the Universe, but physicists find that it is created by many types of nuclear reaction.

7 The Standard Model

Collectively known as 'fermions', the twelve matter particles (six quarks and six leptons) form the basis of the Standard Model of particle physics, alongside a group of force-carrying particles called 'bosons'. Fermions and bosons are distinguished by their different spins (a quantum property that can be imagined as similar to the angular momentum of a spinning top, though it's not quite 'real' in the same way). Spin comes in quantized units: the fermions of the Standard Model all have spin of ½, while bosons have whole-number spins: 1 in the case of the boson particles that carry forces between matter particles, and 0 in the case of the elusive Higgs boson (in case you thought we'd forgotten!)

8 So where does the Higgs fit in?

The Higgs boson is the outward manifestation of a theory called the 'Higgs effect'. So-called 'gauge theories' of fundamental forces explain how matter particles interact (essentially, fermions exchange 'gauge bosons' to 'tell' each other they're there and how to react). But these theories *don't* explain why gauge bosons exhibit the masses that they do. Around 1964, several teams of scientists proposed the existence of a force field inherent to the structure of spacetime, now known as the 'Higgs field'. Gauge bosons and other particles with the property we know as 'mass' interact with this field, slowing down and releasing Higgs bosons in the process.

9 Effects of the Higgs boson

The Higgs field and Higgs boson get their names from Peter Higgs (b.1929), one of the pioneers behind the theory. The nickname 'God particle' was first applied to the Higgs in a 1993 book by Leon Lederman and Dick Teresi (although a lot of other particle physicists, including Higgs himself, aren't so keen on the term.) Experiments into the behaviour of particles at high energies soon offered support for the idea of a field that affects massive objects. It's important to understand that the Higgs doesn't have mass itself – the Higgs effect can be compared to swimming through treacle, with the bosons as the 'stickiness' factor. Predictions estimated that, if it could be created in isolation, the particle's mass would be greater than all but the heaviest top quark, and that it would probably remain stable for less than one sextillionth of a second.

10 Finding the Higgs

The high mass and instability of the Higgs boson put direct observation beyond the range of current particle accelerator experiments (up to and including the LHC). So how was it discovered? The key was not to look for the boson itself, but for the next generation of particles released by its decay, whose properties could be predicted by theoretical models. Evidence for the Higgs was found through analysis of some 300 trillon proton–proton collisions over the first Higgs experimental run in 2010–11. By 2012, the scientists were sufficiently convinced by the data to announce their discovery of a 'Higgs-like particle' with exactly the properties predicted by theory.

TALK LIKE A GENIUS

‘It’s really hard to come up with a good analogy for the way the Higgs mechanism works – the thing to remember is that the bosons aren’t just “piling on” to the particle, and the field isn’t actually resisting motion. Also, bear in mind that the mechanism was only developed to explain why certain force-carrying particles have mass – it’s not responsible for most of the mass in normal matter particles.’

‘Scientists tend to be cautious about big discoveries like this – and with good reason. Although the particle discovered in the LHC has the properties they’d expect from a Higgs boson, they’ve not yet got conclusive proof that it plays the role predicted by the Higgs effect. So for the moment it remains a “candidate” Higgs boson – albeit a very convincing one as, otherwise, there’s no good reason for this particular particle to exist. It’ll probably stay that way until we build an even bigger collider.’

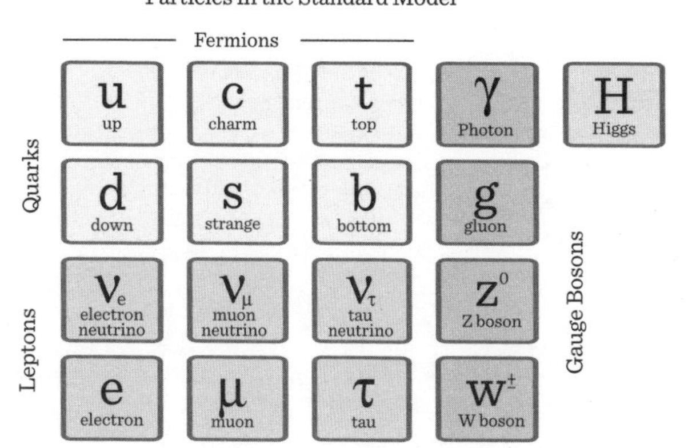

WERE YOU A GENIUS?

1 TRUE – Einstein showed that a phenomenon called ‘Brownian motion’, the vibration of tiny particles suspended in air or liquid, is due to their being jostled by invisible atoms.

2 FALSE – nuclei are actually held together because the attractive ‘strong nuclear force’ is much more powerful than the repulsive electromagnetic force over tiny distances.

3 FALSE – atoms of the same element can vary in their number of neutrons, creating different ‘isotopes’.

4 FALSE – antimatter can also be created around black holes, such as the giant one at the centre of the Milky Way.

5 TRUE – physicists hope that more Higgs particles could help explain the mass of matter particles and even elusive ‘dark matter’.

The Higgs boson explains how force-transmitting particles can change their mass during interactions – but we still don’t know how other particles get mass.

Theories of everything

'Theoretical physicists ... have always believed that the laws of nature are the unique, inevitable consequence of some elegant mathematical principle... The empirical evidence points ... to the opposite conclusion.'

LEONARD SUSSKIND

The quest for a model that successfully unites all aspects of fundamental physics in a single mathematical description has occupied physicists for more than a century, with bestseller lists frequently topped by explanation of the latest possible 'Theory of Everything'. It's clearly something that any self-respecting genius should have an opinion about, but how can you tell your string theory from your loop quantum gravity and pick the electronuclear force out of your branes?

Does the search for a simple theory explaining every fundamental force and subatomic particle actually mean making our Universe *more* complicated?

1 Operating over billions of light years, gravity is the strongest force in the Universe, vastly more powerful than electromagnetism and the weak and strong nuclear forces.

TRUE / FALSE

2 Virtual messenger particles are a neat mathematical device for explaining how particles interact with each other, but so far there's no evidence that they actually exist.

TRUE / FALSE

3 A superstring is a variety of energy wave that some theories say manifests in different types of particles depending on how the string vibrates.

TRUE / FALSE

4 Current versions of string theory require the existence of a total of 26 dimensions.

TRUE / FALSE

5 If we were much, much smaller, then we could perceive the existence of extra dimensions in the Universe.

TRUE / FALSE

TEN THINGS A GENIUS KNOWS

1 The four forces

In the previous chapter we saw how all the matter in the Universe can be broken down into a dozen elementary particles called fermions (and further split into heavyweight quarks and lightweight leptons). But how do these particles interact with each other to join together and form larger objects? Scientists have long been aware of two distinct forces: gravity exerted by objects with mass, and electromagnetic attraction or repulsion between objects with electric charge or magnetic polarity. In the mid-20th century, investigations of atomic nuclei revealed two more forces at work on the smallest scales. Known as the 'strong' and 'weak' nuclear forces, both are much more powerful than either gravity or electromagnetism, but only operate over extremely short ranges.

2 Quantum electrodynamics

According to the Standard Model of particle physics, forces can be explained by the action of force fields that permeate spacetime and affect matter particles differently depending on their properties. The most famous of these theories, 'Quantum Electrodynamics' or QED, was developed by Richard Feynman and others in the 1940s, and describes the way that electromagnetic force works. Like other field theories, it involves the exchange of messenger particles called 'gauge bosons' – bursts of electromagnetic energy that make susceptible particles aware of each others' presence and tell them how they should react. In the case of QED, the gauge bosons are none other than photons, massless quanta of light and other forms of electromagnetic radiation.

3 Virtual messengers

An important difference between the photons in QED and those measured in other situations is that photons acting as gauge bosons don't have to be 'real' in the sense that we'd normally understand. Heisenberg's uncertainty principle, one of the keystones of quantum theory, permits minute fluctuations in the energy of the empty space over short periods, allowing 'virtual' particle–antiparticle pairs to pop into existence and later annihilate with each other. Strange though it sounds, this is how particles in QED and other field theories send messages to each other. What's more, because the messenger photons are massless, they only require a tiny amount of energy and can therefore persist for long periods, explaining the long range of the electromagnetic force.

4 Colour theory

QED was so successful at explaining electromagnetic interactions that physicists naturally pursued similar gauge theories to explain the other subatomic forces. The strong force (which binds the protons and neutrons of the atomic nucleus, and bonds the individual quarks inside *them*) is described by quantum chromodynamics (QCD), a theory that ascribes each quark with a 'colour charge' (colour in this case being a neat analogy, rather than the familiar everyday property). QCD shows how certain colours can be mixed in pairs or triplets, like paints on a palette, to form neutral combinations. The strong force is carried by two types of messenger – virtual gauge bosons called 'gluons' exchanged between individual quarks, and heavier virtual particles called 'pions' (actually paired quarks) between protons and neutrons. The higher mass of the particles involved limits how long their energy can be 'borrowed' and therefore constrains the range of the force.

5 The weird weak force

The weak nuclear force (only weak by comparison with the strong force) is important because it allows quarks to change from one flavour to another (from up to down, or vice versa) – an important process in radioactive decay. Depending on the precise interaction occurring, it can be transmitted by any one of three separate gauge bosons, the electrically charged W^+ and W^-, and the neutral Z^0, and it can be described by a field theory of its own called 'quantum flavour dynamics' (QFD). However, it's more often described in terms of the 'electroweak theory', a mathematical model that describes how the weak and electromagnetic forces behave in the same way in extreme environments such as those found in particle accelerators.

6 Unifying the forces

Formulated in the 1970s, the electroweak theory was an important first step towards unifying the various fundamental forces in a single description. In theory, the strong force should be relatively 'easy' to merge into the electroweak theory at high enough energies, giving rise to a so-called 'Grand Unified Theory' (GUT). Various GUTs have been proposed, each making distinct predictions that may one day allow one or another to be confirmed – but what about gravitation? The best description for this force comes not from a gauge theory, but from Einstein's theory of general relativity. Finding a working quantum description for gravity that meshes with the other forces is a huge challenge, but such a description really would be deserving of the title 'Theory of Everything'.

7 Searching for quantum gravity

Why is gravity such a problem? The main issue is that it behaves very differently from other forces: gravity is a property of matter in bulk, so weak as to be undetectable on the quantum scale of individual atoms, yet with such range that it makes itself felt over larger distances than even electromagnetic fields. Reconciling these properties with a gauge theory would be tough enough without the existence of a perfectly good large-scale model of gravity in the form of general relativity, and while physicists have hypothesized a gravity-carrying boson called the 'graviton', and predicted certain situations in which we might detect it, there's no evidence for its existence so far.

8 String theories

As if unifying the fundamental forces was not ambitious enough, theoretical physicists also believe that a good Theory of Everything should naturally give rise to the various properties of both fermions (matter particles) and bosons (force-carrying particles). The best-known attempts to explain these properties so far are 'string theories', which envisage particles as tiny strings or loops of energy vibrating in various frequencies: the strings naturally fall into harmonic patterns at certain frequencies (similar to the harmonics of a violin string), corresponding to specific values of the various properties. This explains why the properties take on discrete quantized values rather than varying continuously. The major challenge with string theories, however, is that they require strings to vibrate in at least ten dimensions – six more than those we can actually see in the Universe around us.

9 Higher dimensions and branes

Supporters of string theory believe that its extra dimensions are other 'directions' in space, at right angles to the three dimensions that we normally encounter. They explain the apparent invisibility of these dimensions by suggesting they are 'compactified', curled up on incredibly small scales (rather like a tangled ball of string that dwindles to a dot when viewed from a great distance). 'M-theory' (the most promising current form of string theory, developed in the 1990s) adds a seventh extra space dimension: unlike the others, this is *uncurled*, creating a 'bulk' volume that separates sheet-like 'branes' of spacetime from each other. According to this theory, our Universe (with its four dimensions of spacetime and six compactified dimensions) is just one brane among many.

10 Alternative Theories of Everything

String theories are promising, but they're not the only game in town. Other potential Theories of Everything get around the awkward need for extra dimensions by instead proposing a different model of spacetime itself: while string theories and general relativity treat spacetime as a smooth continuum, rival theories suggest there are such things as 'discrete quanta' (smallest possible units) in each dimension. Models with names like 'loop quantum gravity' (LQG) and 'causal set theory' suggest that if we abandon our preconception of continuous spacetime, we will have a better chance of developing a working theory of quantum gravity without the need for extra dimensions, strings or other bizarre new concepts.

TALK LIKE A GENIUS

❛ Part of the reason that QED is so well known is because Feynman came up with a simple graphical way to represent it. "Feynman diagrams" show the paths of matter particles in straight lines and bosons as wavy ones. Interactions happen at the junctions between lines, and you can do an operation called a "path integral" that shows the simplest interactions are the most likely to happen. ❜

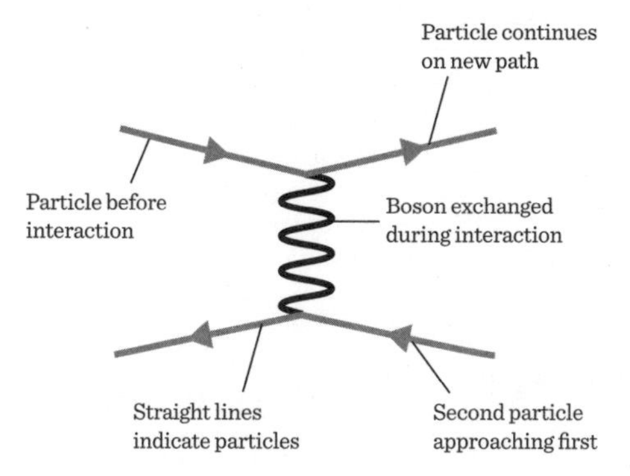

❛ Basically, there are two approaches to quantum gravity: string theory starts with the "quantum" bit and then tries to find a way for gravity to work in the same way producing large-scale effects that look like general relativity. LQG and similar theories start with general relativity and try to figure out how that would operate on a quantum scale, and what effect that would have on the other quantum forces. ❜

WERE YOU A GENIUS?

1 FALSE – strictly speaking, gravity is the *weakest* force of all, since it exerts only a tiny force between individual particles.

2 FALSE –a phenomenon called the 'Casimir effect', which pushes together parallel metal plates in a vacuum, provides real evidence for virtual particles.

3 TRUE – superstrings are used in 'supersymmetric' theories (models that pair each known boson and fermion with an undiscovered counterpart of the opposite type).

4 FALSE – the addition of extra supersymmetric particles in the 1980s reduced the number of required dimensions to a more manageable ten.

5 TRUE – string theories assume the extra dimensions are 'curled up' on tiny scales, far smaller than even the tiniest subatomic particles.

With two of the four forces successfully unified and high hopes of bringing a third into the fold, gravity remains the stubborn outsider.

The Big Bang

'A universe that came from nothing in the Big Bang will disappear into nothing at the big crunch. Its glorious few zillion years of existence not even a memory.'

PAUL DAVIES

The Big Bang theory argues that our Universe began in a vast explosion some 13.8 billion years ago, to which the rest of cosmic history is more or less a footnote. It's by far the most successful scientific attempt at explaining the origins of our Universe, describing many of its features from the expansion of space to the distribution of elements. But it's not without its problems – most frustratingly around the elusive events that triggered the explosion in the first place, and the fate of the cosmos in the far distant future.

If the Universe is expanding from a moment of creation, what does this tell us about its past, present and future evolution?

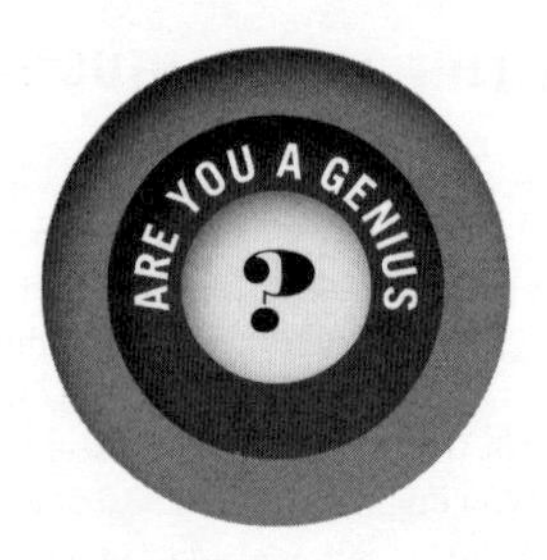

1 Astronomers calculate the distance to other galaxies by measuring the pulsations of stars within them.

TRUE / FALSE

2 The discovery that the light of more distant galaxies is redder than that of nearby ones led astronomers to conclude that the Universe is expanding.

TRUE / FALSE

3 In 1927, Georges Lemaître became the first astronomer to argue that cosmic expansion implied the Universe began in a 'Big Bang' explosion.

TRUE / FALSE

4 A substantial amount of the Universe's unseen 'dark matter' could be made up of black holes, burnt-out stars, planets and other unseen, but 'normal', objects.

TRUE / FALSE

5 The present-day structure of the Universe probably originates from clumps of dark matter that began to coalesce shortly after the Big Bang.

TRUE / FALSE

TEN THINGS A GENIUS KNOWS

1 Expanding space

The Big Bang theory originates with a pretty simple observation: namely, that whatever direction we look in the Universe, distant galaxies are moving away from us, and the further away a galaxy is, the *faster* that motion tends to be. At face value, you might think this means we're uniquely unpopular for some reason, but a better way of interpreting this cosmic expansion is to assume that the Universe as a whole is expanding, with galaxies moving away from each other as the space between them stretches, rather like raisins in a rising cake mix. If space is expanding, then it stands to reason things were closer together in the past – and ultimately jammed together at a single point in space: the Big Bang.

2 Hubble's discoveries

The key evidence for the Big Bang comes from the work of Edwin Hubble (1889–1953; he of Space Telescope fame). Until less than a century ago, most astronomers assumed that the Milky Way galaxy, a system of many billions of stars including our solar, was the entirety of the cosmos. However, a few argued that fuzzy star clouds called 'spiral nebulae' were distant galaxies in their own right. In the 1920s, Hubble worked out the intrinsic brightness of certain stars within the spiral nebulae, showing they were so far away that their light takes millions of years to reach us. As an encore, he measured shifts in the colour of light from each galaxy that indicated their speed of motion towards or away from Earth. This revealed 'Hubble's Law', the link between galaxy distance and speed of recession.

3 An origin for the Universe

Before Hubble's work, most scientists had spent the previous century getting to grips with the idea that the Universe had no beginning. Evidence from geology and evolution suggested that the Earth was countless millions of years old, clearly contradicting the Bible, so perhaps the cosmos had always existed? When Einstein formulated his general theory of relativity in 1915, he deliberately included a 'cosmological constant' that would counteract the Universe's tendency to collapse under its own weight, thinking it was needed to produce a 'steady state' Universe with an infinite lifetime. Even today, some astronomers are uneasy with the idea of the Universe having a distinct origin in time.

4 The primeval atom

Belgian priest and astronomer Georges Lemaître, who predicted cosmic expansion before Hubble found the evidence for it, was also the first person to follow its implications to their logical conclusion. Since the Universe as a whole should behave just like gas in a piston, the earlier, denser cosmos would necessarily also be hotter. This led Lemaître to hypothesize that everything emerged from a superdense, superhot 'primeval atom'. Figures for the rate of expansion, however, were so loosely defined that the age of the Universe varied wildly: it took the work of the Hubble Space Telescope to pin down the expansion, and date the convergence to about 13.8 billion years ago.

5 The afterglow of creation

Clinching proof of the Big Bang comes from another important discovery, the Cosmic Microwave Background Radiation (CMBR). As early as the 1940s, astronomers realized that the hot conditions of the early Universe should have produced intense radiation that we might still detect, reaching us today after billions of years of travel from the very edges of the cosmos. During that long journey, the expansion of space has stretched the wavelengths of the radiation and robbed it of much of its energy, so it reaches Earth as a weak microwave signal. These microwaves, heating the whole sky to 2.7°C (4.9 °F) above absolute zero, were detected in 1964, and mapping tiny variations in their wavelengths and other properties offers important clues to conditions shortly after the Big Bang.

6 Forging the elements

Conditions in the early Universe were so extreme that matter took on very different forms from today. Winding the clock back, energies would have outstripped those in particle accelerators like the Large Hadron Collider (LHC), with individual atoms broken down into separate element particles. In the very first seconds of creation, matter and energy were interchangeable according to Einstein's $E=mc^2$. High-mass particles, such as quarks, could only be formed for a brief moment and rapidly came together to form protons and neutrons, which themselves bonded to form the simplest atomic nuclei (hydrogen, helium and lithium) in a process called 'nucleosynthesis'. Less massive electrons were produced for several minutes, and didn't bind together with the nuclei until cosmic temperatures had dropped below 3,000°C (5,400°F), about 380,000 years after the Big Bang itself.

7 Cosmic inflation

All those unbound electrons in the early Universe meant that space was foggy – photons of radiation ricocheted back and forth between densely packed particles and pushed them apart, which should have kept the density more or less uniform up until the 'recombination era' 380,000 years ago. So how did today's distinctly uneven, large-scale cosmic structure, with huge galaxy clusters around apparently empty voids, develop from such smooth beginnings? In the early 1980s, cosmologist Alan Guth and others came up with an ingenious 'patch' for the theory – an idea called 'inflation' that suggests the Universe we can see today developed from a tiny region of the primordial Big Bang, blown up in a sudden and violent release of energy in the first fraction of a second of cosmic history. Inflation suggests that today's structure is an echo of quantum-scale variations in the infant Universe, and the discovery of tiny irregularities already present in the CMBR (which originates from the recombination era) backs up the idea.

8 Dark matter

Since the 1970s, cosmologists have come to recognize that visible matter tells a misleading story about the Universe – all the material present in stars, planets and interstellar gas and dust clouds is vastly outweighed by mysterious 'dark matter' that only gives itself away through its gravitational influence. Dark matter is probably made up of exotic particles that are not only non-luminous themselves, but are also transparent to light. While the Big Bang theory does have to be rejigged a bit to account for this mystery matter, in some ways dark matter solves more problems than it raises, not least because dark matter clumps can provide 'seeds' for galaxy formation in the early Universe.

9 Dark energy

In the 1990s, astronomers surveyed distant galaxies looking for brilliant exploding stars whose brightness could be used to cross-check measurements of Hubble's law in the nearby Universe. They expected to find signs that the Universe was either expanding at a uniform rate, or slowing down, thanks to the gravitational pull of visible and dark matter within it. But instead, they discovered that cosmic expansion is *speeding up* – a result now attributed to a mysterious but powerful force called 'dark energy'. Nobody knows quite what dark energy is, but it may be a 'cosmological constant' driving the expansion of space, similar to the one proposed and later abandoned by Einstein.

10 The fate of the cosmos

Studying the Big Bang not only tells us about the past and present of the Universe, but also about its future. The precise speed of cosmic expansion and the amounts of matter and dark energy in the cosmos determine whether gravity will eventually slow and reverse the expansion, pulling everything back to a 'big crunch', or whether expansion will continue forever, with the Universe slowly growing colder and darker in a 'big chill' as the raw elements that make stars shine are used up and scattered across space. Another scenario, suggested by recent evidence that the strength of dark energy is increasing over time, sees expansion accelerate until matter is torn apart in a 'big rip'. Fortunately, we've probably got a few trillion years before any of these situations become pressing.

TALK LIKE A GENIUS

“The last 500 years of astronomy have seen us go from the centre of the Universe, to one planet orbiting the Sun, to a single solar system halfway across the Milky Way galaxy. And it turns out there are at least as many galaxies in the Universe as there are stars in our galaxy. If that tells us one thing, it's the Copernican principle – never assume there's anything special about our place in the Universe.”

“It's ironic that we call it the “Big Bang”, because that term was coined as an insult by Fred Hoyle, one of the theory's strongest opponents.”

“In its purest form, the theory is quite a clever dodge – you can't talk about what happened *before* the Big Bang because the explosion didn't just create matter, it created time and space themselves. And you can't even talk about what happened in that first tiny instant because the Universe was so small and dense that quantum uncertainty takes over and the normal laws of physics don't apply. These days, of course, there are also plenty of multiverse theories that do offer possible explanations for what caused the Big Bang...”

WERE YOU A GENIUS?

1 TRUE – some stars vary in regular cycles linked to their intrinsic luminosity. By measuring the period of variation, astronomers can work out their true luminosity and distance.

2 TRUE – this ‘red shift’ is caused both by the motion of galaxies away from Earth, and by the stretching of light as it crosses expanding space.

3 TRUE –however, Alexander Friedmann was the first to show that an expanding Universe fitted with general relativity, as early as 1922.

4 FALSE – attempts to measure the distribution of dense dark objects in our galaxy suggest they're just not widespread enough to contribute much to explaining the dark matter problem.

5 TRUE – because dark matter is immune to interactions with light, it could begin to clump together through gravity very early on.

The Big Bang theory explains everything about where the Universe came from, except for the tricky bit at the very beginning.

Special and general relativity

'Spacetime tells matter how to move; matter tells spacetime how to curve.'

JOHN ARCHIBALD WHEELER

Einstein's theories of special and general relativity, published in 1905 and 1915, form much of the bedrock of modern physics, and are precisely the sort of thing any self-respecting genius should know about. It's easy to get lost in the analogies about fast-moving clocks and rubber sheets, but the important things to understand are that the speed of light in a vacuum is constant (and nothing else can reach that speed, so the rest of physics just has to bend to accommodate it), and that gravity and acceleration are essentially the same.

Relativity turns our everyday experience of physics on its head, but it's still by far the best description of the way the Universe really works – so what's it all about?

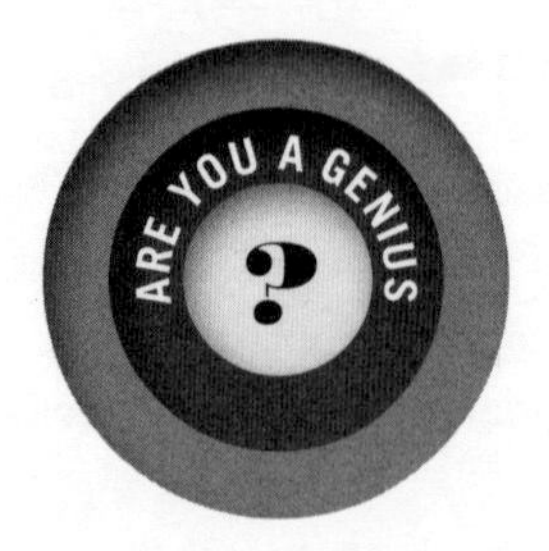

1 The principle of relativity was first proposed by Albert Einstein in 1905.

TRUE / FALSE

2 Einstein was something of an academic failure until he published his four revolutionary papers in 1905.

TRUE / FALSE

3 Einstein's model of general relativity was entirely based on 'thought experiments', and it was not until 1919 that astronomers found the first evidence to support it.

TRUE / FALSE

4 Einstein's 'twin paradox' suggests that if a twin goes on a long voyage at close to light speed, they will age more slowly than their sibling who stays at home on Earth, even though both twins will see time pass more slowly for their sibling because of their relative motion.

TRUE / FALSE

5 General relativity was only finally accepted when astronomers confirmed the detection of gravitational waves from colliding black holes in 2016.

TRUE / FALSE

TEN THINGS A GENIUS KNOWS

1 The speed of light

Einstein's theory emerged from efforts to resolve a paradox in the behaviour of light – namely the fact that it appears to be fixed, regardless of the relative motion of a source of light and its observer. That's a stark contrast to everyday physics (where speeds usually add up – imagine catching a football thrown from a moving train), but it took a long time to recognize, because the speed of light is so fast that any other relative motion is usually only going to make a tiny impression. It was only in the mid-19th century, when James Clerk Maxwell showed how his electromagnetic theory of light *always* produced waves with a speed of 299,792 km/s (186,282 miles per sec, denoted c for short) that physicists began to suspect something was up.

2 The search for the aether

Maxwell's model of light as a wave raised the obvious question as to what medium that wave was travelling through, and led physicists to imagine an all-pervading 'luminiferous aether'. The relativity revolution really got started when US scientists Albert A. Michelson and Edward Morley came up with an ingenious new experiment that should have detected the tiny difference in the speed of light caused by Earth's motion through the aether. When Michelson and Morley drew a blank in 1889, physics was thrown into something of a crisis. Various ingenious fudges were suggested, but Einstein was one of the few to take the result at face value, and the only one to realize it meant the laws of physics required rewriting from the ground up.

3 Special relativity

The long-standing principle of relativity is simply the idea that laws of physics should behave in the same way for all experimenters, regardless of their 'frame of reference'. So, a scientist set up in a laboratory on a moving railway carriage should detect identical physics to a colleague in the signal box. 'Special' refers to a limitation on Einstein's 1905 paper that simplifies things by only considering 'inertial' laboratories (those that are subject to identical forces and are therefore not accelerating or decelerating relative to each other). Einstein considered what would happen if one experimenter was in a stationary laboratory while the others moved at a speed close to c – in order for them both to measure the same fixed speed of light, what else must they see differently?

4 Relativistic effects

Einstein identified a handful of effects that would be measured differently between frames of reference in relativistic motion (relative speeds comparable to the speed of light). One is the length of objects: in a 'moving' laboratory a metre-rule should appear shorter as viewed from outside (the so-called 'Lorentz contraction'). Another is the flow of time, which slows down or 'dilates' inside the moving laboratory as measured from outside. In both cases, the experimenter inside the moving lab will detect nothing strange themselves. That might make these phenomena sound like illusions, but in fact they're nothing of the sort. The principle of relativity means that the 'stationary' observer cannot have a privileged position from which only they can observe the 'real' physics, so if they carry out the same experiments, then the same distortions will be clear to the observer watching from the moving lab.

5 $E=mc^2$

Another important aspect of special relativity has implications that extend far beyond relativistic motion. Einstein wondered what would happen if you kept on using energy to accelerate a body already close to the speed of light. Since the body could never reach the speed of light itself, Einstein realized the additional energy would instead be increasingly diverted to increasing its mass: that way the body's energy and momentum can continue to increase even when speed cannot. Therefore, Einstein realized, mass and energy are basically equivalent (and, in various scenarios, interchangeable), with a body's energy E linked to its mass m by the famous equation $E=mc^2$.

6 Spacetime

Everyday experience suggests that the three dimensions of space are fixed at right angles to each other. Time, meanwhile, seems to be something else entirely, flowing in a unique 'direction' of its own. Special relativity, however, shows that all four dimensions are linked and can be traded off against each other. Einstein's one-time tutor Hermann Minkowski developed a geometrical approach to 'spacetime', which offers one of the easiest ways to grasp the theory: each 'frame of reference' can be thought of as a region with perpendicular dimensions, but relative motion near the speed of light causes frames of reference to rotate in relation to each other so that certain dimensions are extended or compressed.

7 General relativity

As its name suggests, general relativity attempts to describe relativity in *all* reference frames (accelerating as well as inertial ones). Einstein's key breakthrough here was to realize that a gravitational field is indistinguishable from a constant accelerating force: in other words, an experimenter on a spaceship undergoing constant acceleration should measure exactly the same physics as one sitting on a planet with a gravitational field. From this, Einstein developed the 'field equations' of general relativity, showing that the presence of large masses can warp spacetime in similar ways to relativistic motion.

8 Warped spacetime

One popular way of visualizing the effects of general relativity is to imagine the spacelike dimensions of spacetime reduced to a two-dimensional 'rubber sheet' that is warped in a third 'timelike' dimension by the presence of a massive object (a planet or star), so that other objects passing through the object's 'gravitational well' have their paths deflected. In this model, the orbit of a planet or moon is simply a path that it follows on the inward slope of the well, moving fast enough to avoid falling inwards. Another way of thinking about the effect is to visualize masses creating hourglass-like 'pinches' in the otherwise orderly grid of three space dimensions.

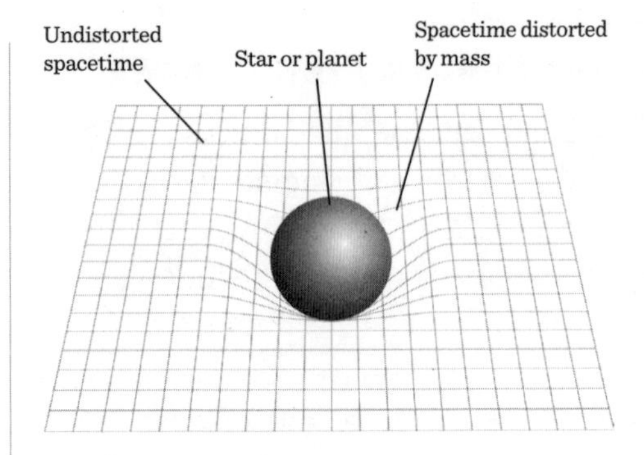

9 Time travel

Einstein's theories have been verified in many different ways: not only through the results of abstruse physics experiments, but in some startling real-world effects. Perhaps the most famous of these is 'gravitational lensing' – the deflection of light passing close to massive objects. Astronomers detected lensing of starlight passing around the Sun during a total solar eclipse as early as 1919. More recently, they have used the lensing effect of distant galaxy clusters to map the distribution of so-called 'dark matter' within them. Closer to home, physicists have shown the effects of time dilation by flying hyper-accurate atomic clocks in supersonic jets and comparing them to their counterparts on the ground.

10 The cosmic speed limit

So can anything travel faster than light? A century of success for Einstein's theory answers a resounding 'no' (although it's always foolish to entirely rule out new physics around the corner). But that doesn't stop scientists and engineers from theorizing possible ways around the limit, ranging from hypothetical 'tachyons' (a mirror-universe of particles that can never travel slower than light-speed) to the strange effect known as 'quantum entanglement'. For anyone planning to explore space beyond our solar system, meanwhile, the light-speed barrier presents a significant challenge. However, time dilation offers something of an escape clause, since time would pass more slowly for the crew of a relativistic spaceship than for those left on Earth.

TALK LIKE A GENIUS

❛ Einstein's so-called *annus mirabilis* in 1905 leaves modern physicists in the dust: not only did he publish papers laying out the theory of special relativity and the idea that mass and energy are the same, but he also produced the first observational evidence for the existence of atoms, and kick-started quantum theory by showing that light travels in photons. Not bad for a patent clerk! ❜

❛ The golden rule of light-speed travel is really that you can't send *information* faster than light – if you could, you'd break the rules of cause and effect and then where would you be? ❜

❛ Several clever people have come up with theoretical ideas for "warp drives" that get around the light-speed barrier. The general idea is that while you can't move a spaceship at a speed faster than *c*, you could create a bubble of normal spacetime around the ship, and then move *that* through its surroundings faster than light speed. It's purely hypothetical at the moment, though, and most physicists suspect there are undiscovered problems that prevent this kind of cheat. ❜

WERE YOU A GENIUS?

1 FALSE –the principle was actually put forward by Galileo in 1632, to counter arguments that Earth's motion would cause obvious effects.

2 TRUE – although Einstein graduated from Zurich with high grades, he was not encouraged to continue his academic career.

3 FALSE – as early as 1915, Einstein showed that general relativity explains a wobble in the orbit of Mercury that classical physics cannot.

4 TRUE – the 'paradox' of one twin returning older than the other is explained by the travelling twin's experience of acceleration and deceleration.

5 FALSE – countless tests had already shown that general relativity was accurate.

Special relativity explains why physics gets weird when objects travel close to the speed of light; general relativity explains why it gets weird in the presence of massive objects.

Black holes

'It is said that fact is sometimes stranger than fiction, and nowhere is this more true than in the case of black holes.'

STEPHEN HAWKING

As some of the most fascinating and mysterious objects in the Universe, black holes are classic genius territory. The concept of an object with such strong gravity that not even light can escape it has a surprisingly long history. But astronomers only began to identify and study black holes in the past few decades – and it turns out they're even more varied and interesting than anyone had suspected.

Close to a black hole, matter is torn to shreds and the laws of physics stop making sense – are you smart enough to navigate this mysterious realm?

1 If you looked outwards while falling into a black hole, you'd be able to see into the future.

TRUE / FALSE

2 Nothing can follow a stable orbit around a black hole without getting sucked in.

TRUE / FALSE

3 It's unlikely ever to happen, but if an accelerator like the Large Hadron Collider (LHC) ever created a microscopic black hole, it could suck in the entire planet in a matter of seconds.

TRUE / FALSE

4 Some physicists think that the Big Bang itself might have left the Universe riddled with microscopic black holes.

TRUE / FALSE

5 There could be a black hole on our cosmic doorstep heading for the solar system and we wouldn't know about it until it was too late.

TRUE / FALSE

TEN THINGS A GENIUS KNOWS

1 What is a black hole?

In essence, a black hole is just an object with an escape velocity higher than the speed of light. This means that nothing can escape its gravity (escape velocity is the minimum speed that an object must travel at, in order to climb out of a gravitational field without ever being slowed to a halt). Earth's escape velocity, which is important if you want to launch an interplanetary spacecraft, is a mere *11.2 kilometres per second,* while that of a black hole is *300,000 km/s*. However, the physics of high-density matter, and the effects of Einstein's relativity mean that black holes are much weirder than the simple definition might suggest.

2 Why a black hole is black

English astronomer John Michell was the first person to suggest black holes might exist, as early as 1783. Based on the then-current theory that light came in small packets called 'corpuscles', he reasoned the possibility of 'dark stars' with gravity so strong that they trapped their own light. Despite several changes in our understanding of light since then, it turns out that general relativity can produce the same effect. Karl Schwarzschild showed in 1915 that, where matter is sufficiently compressed, equations of general relativity can produce 'singularities' where the laws of physics break down. A few years later, Arthur Eddington showed what really happens to light around a black hole: the speed of light itself must remain constant and cannot actually be slowed down, so instead light has its wavelengths stretched or 'red shifted' to infinite, undetectable wavelengths.

3 Finding black holes

So how do you detect a distant object that, by definition, doesn't allow light to escape from its surface? The answer is to look for signs of its influence on nearby objects. In particular, astronomers focus on sources of high-energy X-rays. Most of these are galaxy-sized clouds of gas that have nothing to do with black holes, but others are compact, starlike, and locked in orbit with a visible star in so-called 'X-ray binary' systems. Astronomers believe such systems are created when a black hole pulls matter away from a companion star and surrounds itself with a superhot X-ray-emitting disc of infalling gas in the process. By measuring how the unseen objects pull their companion stars around, astronomers can prove that some of them are so massive and dense that they must be black holes.

4 Where do black holes come from?

Astronomers think that the vast majority of black holes in the Universe are created by the death of massive stars. Such stars generate heavy elements in their cores through the nuclear fusion process that makes them shine, but when this process is finally exhausted, the core collapses with huge force, generating a shockwave that incinerates the star's outer layers in a brilliant supernova explosion. The force of collapse breaks atoms in the core apart and creates a dense soup of neutrons, and, in most supernovae, pressure between neutrons eventually brings the collapse to a halt (creating a neutron star that typically compresses the mass of our Sun into a city-sized sphere). However, if the collapsing core has between two and three times the Sun's mass, the forces involved can break neutrons apart and collapse continues until a black hole is formed.

5 Event horizons

Although featureless from the outside, a black hole actually has a well-defined structure. Its mass is concentrated in a single point called the 'singularity' at the very centre, but its visible outer edge, the 'event horizon' is defined by the point at which the escape velocity exceeds the speed of light. The event horizon may be a perfect sphere but since, in reality, most black holes probably spin very rapidly (inheriting their rotation from their parent stars), it is more likely to bulge outwards around its 'equator'. Seen from nearby space, a lone black hole would be surrounded by a curious halo created where light rays from more distant stars have their paths deflected.

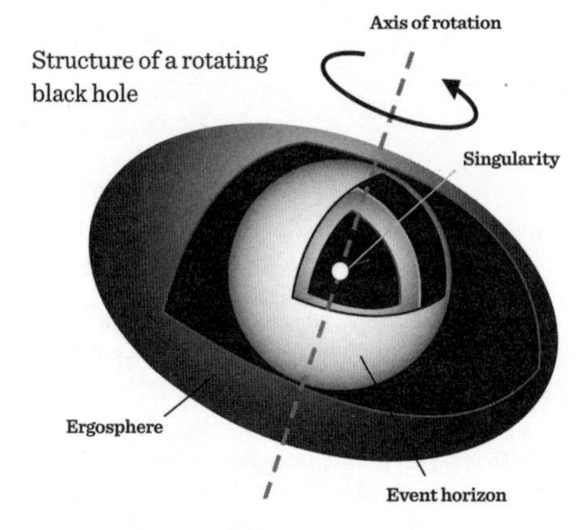

6 Spaghettification

The extreme conditions around black holes push the laws of physics to their limits. As material falls towards the event horizon, the pull of gravity increases rapidly, creating tidal forces that attract one side of an object more than the other. Ultimately, these forces tear objects apart in a process called 'spaghettification'. At the same time, the warping of spacetime creates a time dilation effect similar to that of special relativity – the flow of time for objects near the event horizon seems to slow down and they seem to hang forever on the edge of the event horizon. By the time an object has vanished for an outside observer, it has long since been consumed within its own frame of reference, ultimately breaking down to subatomic particles and merging with the singularity.

7 Supermassive black holes

While most black holes have starlike masses, some are monsters with a mass of millions of Suns. These supermassive black holes lurk in the hearts of most (perhaps all) galaxies, and form the kernel around which huge numbers of stars orbit to form the central core of a galaxy. Our Milky Way galaxy's own central black hole (some 26,000 light years from Earth) has a mass of 4.1 million Suns (as calculated from the speedy orbits of nearby stars). Monster black holes probably began life as the cores of gargantuan stars in the early Universe, but grew rapidly by merging together and gobbling up nearby gas, dust and other stars.

8 Quasars

Most supermassive black holes (including the Milky Way's) are dormant; stars and other material orbit at a safe distance and so they are starved of fuel. A fair number of galaxies, however, have black holes that are actively feeding, pumping out high-energy X-rays and other radiation as they pull clouds of shredded matter to their doom. Radiation from around the black hole can outshine the light of the surrounding galaxy, and varies rapidly. The brightest of these objects, called 'quasars', can be seen across billions of light years of space, making them some of the most distant (and also, therefore, earliest) objects in the known Universe. All galaxies are thought to go through a quasar stage during their formation, and many are reignited when galaxies collide.

9 Hawking radiation

Celebrated physicist Stephen Hawking made his reputation through his work on black holes – particularly his discovery that black holes *do* emit radiation of sorts and therefore lose energy and 'evaporate' over time. 'Hawking radiation' arises when pairs of virtual particles (see page 174) pop into existence close to the event horizon. If one particle is swallowed by the black hole before it can reunite with its counterpart, the other is forced to become 'real', by borrowing energy and mass from the black hole. Depending on its mass, a black hole might take trillions of years to evaporate, eventually coming apart in a burst of radiation as its mass drops below the critical threshold.

10 Wormholes

In terms of general relativity, black holes form very deep, inescapable 'gravitational wells' in spacetime. In theory, it might, therefore, be possible for such a black hole to punch its way through to a completely different part of spacetime where it could link up with another one. The resulting tunnel, known as an Einstein–Rosen Bridge or 'wormhole', could potentially offer a shortcut across the cosmos. However, a wormhole could only be navigated if some way could be found of avoiding the central singularity, which puts it firmly in the realm of science fiction – for now.

TALK LIKE A GENIUS

❛ Actually, Einstein's field equations allow for the possibility of four different types of black hole, depending on whether or not the singularity has an electric charge, and on whether it's rotating or not. In practice, most black holes probably don't carry electric charge, but on the other hand they do probably rotate. ❜

❛ New Zealander Roy Kerr worked out that a spinning black hole has a region outside of its event horizon called the "ergosphere" where spacetime is being twisted. Roger Penrose showed that an object falling into the ergosphere can pick up enough energy to escape before it's dragged over the event horizon, creating huge bursts of energy that gradually drain away the black hole's angular momentum and slow its rotation. ❜

❛ The term "wormhole" was coined in 1957, but only really became a hot topic after Carl Sagan asked a friend, theoretical physicist Kip Thorne, to come up with a plausible means of faster-than-light travel to use in his 1985 sci-fi novel *Contact*. Thorne was also a consultant on the physics and appearance of the black hole in Christopher Nolan's *Interstellar*. ❜

WERE YOU A GENIUS?

1 FALSE – the passage of time in the outside Universe would indeed appear to run faster, but your acceleration would stretch the light on its way to you into invisibility.

2 FALSE – a black hole is just like any other massive object, so provided you move fast enough, you can orbit without falling in.

3 FALSE – the gravity of a micro-black hole would be so weak that it could only grow by swallowing particles that stray into its path.

4 TRUE – Stephen Hawking proposed that the Big Bang could have seeded the Universe with small black holes as early as 1971, and the theory hasn't yet been disproved.

5 FALSE – we'd have thousands of years of advance warning as an approaching black hole's gravity disrupted the paths of nearby stars.

Black holes create some of the most violent physics in the Universe, and offer laboratories for testing general relativity.

Multiverses

'There's reason to suspect that our Universe may be one of many – a single bubble in a huge bubble bath of other universes.'

BRIAN GREENE

Is our Universe the only one of its kind, or just one among countless others in a multiverse that could be infinite in extent? It's a mind-boggling idea, but it's one that many cosmologists are taking increasingly seriously. So how could a multiverse exist, what are the implications for the past, present and future of our Universe – and does this mean there's another you somewhere out there, reading an identical version of this book printed in blue ink?

Proving the existence of an infinite multiverse would change our view of the cosmos more fundamentally than the discovery that the Earth goes around the Sun.

1 Light from the most distant edge of the visible Universe is noticeably redder than that from nearby space because its waves are stretched by the Doppler effect – the same effect that changes the pitch of a passing ambulance.

TRUE / FALSE

2 If you could somehow keep going in one direction across many neighbouring observable Universes, you'd eventually find yourself approaching our Universe from the opposite direction.

TRUE / FALSE

3 In 2017, astronomers published evidence of a cold spot in the cosmic microwave background radiation that might be the first sign of another multiverse intruding into our own.

TRUE / FALSE

4 The simplest cosmological models suggest that if we could travel across a vast distance of space, we'd eventually find another Earth.

TRUE / FALSE

5 The theory of an infinite multiverse allows for consciousness to spontaneously pop into existence in space.

TRUE / FALSE

TEN THINGS A GENIUS KNOWS

1 The limits of the Universe

The fixed speed of light and finite age of our Universe place a fundamental limit on the distance we can observe in our Universe – no matter how powerful our telescopes, it's impossible for us to see objects whose light has not had time to reach us in the 13.8 billion light years since the Big Bang. What's more, the further away we look back in space, the further back we are looking in time, so we can only see what distant objects looked like in the past, not the present. In effect, this limits the 'observable Universe' around us to a spherical shell about 93 billion light years across (more than 13.8 billion light years, because the Universe has expanded considerably since the most distant light set out on its journey towards us).

2 Beyond the observable Universe

However, there's no reason to think that the edge of our observable Universe is the edge of *everything* – an astronomer sitting on a planet 46.5 billion light years from Earth today would be surrounded by their own observable Universe – if they looked in our direction they might see the raw materials of the Milky Way in the process of coming together, and if they looked in the opposite direction they'd be staring into regions of the cosmos forever hidden from our view. An astronomer on the opposite edge of *that* Universe would have an observable Universe with no overlap to our own, and so on and so on. All these neighbouring overlapping bubbles of spacetime together make up the simplest form of 'multiverse'.

3 The shape of space

In theory, every bubble in this simple multiverse blew up from a small region of the infant Universe that existed before cosmic inflation took hold (see page 190) – but just how many of them are there? Could space carry on forever? The answer to that depends on the shape of space – is it 'flat', carrying on forever in all directions, or could it curve back on itself (warped by the mass of visible and dark matter within it) to form a closed shape? The discovery of dark energy causing spacetime to stretch at an ever-increasing rate suggests the former is most likely – our observable Universe is just one among an infinite number.

4 The four types of multiverse

Theoretical physicist Max Tegmark identifies four distinct levels of multiverse, each weirder and more abstract than the last. Level 1 is simply 'more spacetime' as described above. A Level 2 multiverse consists of bubble Universes like our own that differ in their number of dimensions and other physical properties, but which nevertheless all originate from the same source – spacetime 'foam' (see below). The 'Many Worlds' interpretation of quantum physics (see page 177) gives rise to the Level 3 multiverse, with countless parallel lower-level multiverses branching off from each other in some higher-dimensional space. Finally, the Level 4 multiverse is a mathematical structure incorporating all the lower levels (a classification that suits Tegmark's belief that reality is fundamentally mathematical in nature).

5 The Level 2 multiverse

A theory called 'chaotic inflation', developed by cosmologist Andrei Linde and others shortly after the original inflation theory was proposed, can give rise to a so-called Level 2 multiverse. Linde and his colleagues wondered why one particular bubble of the primordial Universe was selected over all the others for inflation into the current Universe. They suggested an alternative in which inflation is an ongoing process – new Universes are constantly inflating out of the high-energy spacetime foam, rather like bubbles spontaneously forming in a bottle of soda. For some cosmologists an added advantage to this theory is that it avoids having a finite beginning of time – the foam is eternal and the Big Bang is just the moment at which our particular Universe condensed out of it.

6 Varied dimensions

Another aspect of Level 2 multiverses is that they can manifest different numbers of dimensions. If superstring theory is correct in its prediction of at least six extra space dimensions, then there are many different ways of 'hiding' the unseen ones, and no particular reason why all Level 1 multiverses should have just three unfolded dimensions. Eternal inflation could also give rise to bubbles with radically different fundamental constants – many would be snuffed out in their infancy, but others could survive and expand, perhaps even colliding with their neighbours.

7 The evidence for multiverses

If another Level 2 multiverse collided with our own, some cosmologists believe that it would produce evidence that might one day be observed. The walls of the expanding bubbles would be extremely rigid and likely travelling at close to the speed of light, as they eat up the empty vacuum around them. As a bubble impinged on our observable Universe, it would send huge ripples of energy propagating across space in a so-called 'cosmic wake'. Such wakes would have various consequences, although their passage through empty space is invisible, so the best place to look for them is in the CMBR microwave radiation from the very edge of the Universe. Here we might expect to find rings of slightly raised temperature, or other giveaway patterns, but even the most powerful of collisions are predicted to produce traces that would be at the very limits of our current observing capabilities.

8 Levels 3 and 4 multiverses

Tegmark's Level 3 and 4 multiverses are far more complex entities that are hard to get a handle on – even Hugh Everett, originator of the Many Worlds interpretation, didn't offer much insight into how his continually branching multiverse worked in practice to create alternative realities at every possible decision point. In this regard, perhaps Tegmark's Level 4 idea that the Universe is fundamentally mathematical in nature is the best way of handling it – the vast demands of mass and energy in a 'real' multiverse can be neatly sidestepped if we're just talking about different sets of numbers running in some vast cosmic simulation.

9 Explaining our fine-tuned Universe

The existence of a complex multiverse may offer a scientific solution to one of the biggest cosmological mysteries – why are we here at all? The range of possible outcomes of the Big Bang was vast, and the chances of a Universe arising in which the balance of matter and energy, and the strength of various fundamental forces, allowed the appearance of stars, planets and life are astronomically slim compared to all the other, more hostile outcomes. Multiverses of Level 2 and higher address part of the problem by removing our cosmic privilege and allowing all those other Universes to simultaneously exist somewhere else, but that still doesn't explain why we're in *this* Universe.

10 The anthropic principle

Cosmologists address the problem through the so-called 'anthropic principle', an observer-centred view of the Universe that comes in weak and strong flavours. The weak principle states that the Universe is the way it is because we're here to see it – if things had turned out differently, we simply wouldn't be around to enjoy the scenery. The strong form of the idea argues, conversely, that cosmic properties are the way they are for some underlying reason – perhaps even in order to make life possible. What feels like a remarkable relapse to religious and philosophical ideas about the role of humans in the Universe, in fact has some scientific grounding in some interpretations of quantum mechanics – is it even possible for the Universe to collapse from its initial quantum state without an observer here to see it?

TALK LIKE A GENIUS

“One alarming possibility is that the arrangement of dimensions we're familiar with is unstable – our entire Universe could be in a so-called “false vacuum” state. If that's the case, then it means that a new and more stable bubble Universe could pop into existence inside ours, expanding and overwhelming everything at the speed of light – cosmologists call it a “Big Slurp”.”

“The idea of a mathematical multiverse is where cosmology really does start to merge into philosophy. Once you accept the idea that someone *could* come up with a simulation smart enough to pass for reality, the sheer weight of probabilities means they probably have, lots of times – which inevitably means we're more likely to be the product of one of those many simulations than 'real' beings existing in the one-and-only reality.”

“Some cosmologists say a multiverse doesn't really solve the fine-tuning issue at all. For instance Paul Steinhardt, one of the pioneers of the inflation theory, argues that it's just shifting the goalposts; inflation was initially invented as an explanation for certain features of the Universe, but now we have to explain why, out of all possible multiverses, we ended up with one that creates these inflationary bubbles.”

WERE YOU A GENIUS?

1 TRUE – however, the reddening effect is actually a combination of the Doppler effect and the stretching of light waves as they pass through space that is itself stretching.

2 FALSE (probably) – this would be true if spacetime was 'closed', and our Level 1 multiverse was spherical, but dark energy suggests instead that spacetime is 'flat' and infinite.

3 TRUE – however, the chances of the cold spot being real evidence of an impinging multiverse are very small.

4 TRUE – based on the number of atoms in our Universe and their possible arrangements, cosmologists have even put a rough (huge!) figure on the distance.

5 TRUE – so-called 'Boltzmann brains' should be able to arise from random fluctuations in a multiverse that is truly infinite.

Spacetime extends far beyond the limits of what we can see, but it's also possible that our entire cosmos is just one among infinite multiverses.

Life in the Universe

'Two possibilities exist: Either we are alone in the Universe or we are not. Both are equally terrifying.' ARTHUR C. CLARKE

Perhaps the biggest and most intriguing question in modern science is whether we're alone in the Universe. Is anybody out there, and if so, what are they like? The past few decades have seen some enormous leaps forward in our understanding of the requirements for life, and just how widespread suitable conditions are likely to be in our galaxy and others. But when it comes to the questions of whether life has evolved elsewhere, and whether it's likely to be intelligent, we're still in the realm of informed guesswork.

Is life common in the Universe, or are we an incredible fluke? And if there are intelligent aliens out there, should we try to say 'hello'?

1 The Arecibo message sent to the stars in 1974 has been criticized because it contains instructions on how aliens could reach Earth.

TRUE / FALSE

2 Astronomers are fascinated by a star called KIC 8462852, which they speculate could be orbited by a huge alien engineering project.

TRUE / FALSE

3 Most attempts to detect alien intelligence have surveyed the sky in a relatively narrow band of radio wavelengths.

TRUE / FALSE

4 SETI scientists have never found a signal from space that they've not been able to eventually explain as either natural or generated by us.

TRUE / FALSE

5 Astronomers estimate there could be tens of billions of potentially habitable planets in our galaxy alone.

TRUE / FALSE

TEN THINGS A GENIUS KNOWS

1 The requirements for life

Before we start looking for life, it's a good idea to have some understanding of the conditions we think are needed for life to get a foothold on any planet. Fundamentally, astrobiologists think that in order for life to appear, an environment needs liquid water, some kind of energy source, and some simple carbon-based chemicals to get it going. Fortunately, both water and carbon compounds seem to be common in our galaxy (and presumably in the wider Universe), and energy can be provided either by radiation from a star, or by volcanic or geothermal activity from within a planet. So where should we start looking?

2 Life on Mars

Earth's next-door neighbour in the solar system is a mostly dry, cold world today, but the famous Red Planet's dusty surface conceals huge amounts of deep-frozen ice. There's evidence that liquid water once ran freely on the surface and the atmosphere was much thicker and warmer when the Martian orbit was slightly different. Today, it's on the outer edge of our solar system's 'habitable zone', but even now briny liquid water can occasionally flow on the surface. Billions of years ago, however, Mars might have offered an ideal environment for the evolution of simple life, and it might cling on underground today, perhaps releasing the occasional puffs of short-lived methane gas that have puzzled scientists over the past few years.

3 Life on moons

Further out in the solar system, the solar system's two largest planets, Jupiter and Saturn, both have potentially habitable moons in orbit around them. Jupiter's large moon Europa and Saturn's much smaller satellite Enceladus are both made from a mix of rock and ice, and trap a layer of liquid water between an outer icy shell and a 'seabed' dotted with active volcanoes. The moons are warmed from within, probably as their interiors are pummelled by the tides raised by their giant parent planets. Water escaping through cracks in the outer crust creates plumes of icy vapour above each world – potentially an ideal way for future space probes to detect signs of aquatic life in the oceans below.

4 Exoplanets

Looking further afield, the prospects for widespread life in our galaxy have been transformed since the 1990s by the discovery of thousands of 'exoplanets' orbiting other stars. Though the properties of both stars and planets vary wildly from our own Sun and its solar system (and detection methods tend to favour large, Jupiter-like planets), it's still possible to model the habitable zone of these alien stars, and to speculate about the possibility of exomoons around giant planets. Several potentially habitable planets have so far been found – including one in orbit around Proxima Centauri, the closest star to our solar system.

5 Signs of extraterrestrial life

Amazingly, astronomers think we might already be able to detect life even on a planet of another star – if we get lucky with the conditions. If a planet with an atmosphere has an orbit that happens to take it across the face of its star as seen from Earth (an event called a transit), then we can detect not just a dip in the star's overall brightness, but also slight changes to the star's rainbow-like spectrum of light as chemicals in the planetary atmosphere absorb specific wavelengths of starlight. These chemical fingerprints could potentially reveal distinct atmospheric gases linked to the presence of life or other surface activity – such as methane released by microbial life or volcanoes, or even pollutants linked to the presence of an industrial civilization.

6 What would aliens be like?

The best predictions of alien life try to put the universal rules of evolution by natural selection to work, while also considering how many of the features of life on Earth have arisen simply by chance. For instance, the 'bilateral symmetry' shared by most Earth organisms is simply a result of a shared common ancestor in the distant past, but does not (necessarily) confer a particular advantage over other body plans such as the five-fold symmetry of starfish. On the other hand, streamlined fishlike features have evolved independently on at least three separate occasions in Earth history (in fish, prehistoric marine reptiles and modern whales), and hence we might expect life to take on fishlike forms in any watery environment.

7 Would aliens be intelligent?

The possibility of life taking hold on planets and moons with suitable environments is one thing – but would it necessarily give rise to intelligent aliens capable of communicating with us? It's sobering to think that if the entire history of life on Earth was compressed to a single day, modern humans only arose in the last seconds before midnight – there's really no evidence of an evolutionary 'selection pressure' for human intelligence operating until the past few million years. If intelligence is as rare a phenomenon as that suggests, then it might take an awful lot of life-bearing planets for there to be a few technological 'communicating civilizations' coexisting at this point in the Milky Way's history.

8 The odds of intelligent life

In 1961, astronomer Frank Drake (a pioneer in the field of SETI, the Search for Extra-Terrestrial Intelligence) devised an equation as a way of calculating the number of communicating civilizations presently inhabiting the Milky Way. The Drake equation multiplies up the number of habitable planets being born in the Milky Way annually by various other factors such as the probability of life, intelligence, and long-range communication technology (plus avoiding a natural or self-inflicted extinction) to come up with an answer. In reality, however, the equation is more of an aid to thinking than a hard-edged mathematical tool – with just one example of such a civilization to work from (our own), we simply don't have enough information to pin down the probabilities, so plausible predictions can range wildly from one civilization to more than 100 million.

9 Alien messages and artefacts

SETI astronomers search for evidence of alien civilizations using a number of different techniques. Although radio signals might seem like the most obvious way of communicating, broadcasting a strong message over many light years of space would take vast amounts of energy, so aliens would be more likely to deliberately target planets on which they have already seen signs of technological life (which might not necessarily include us, given that our industrial revolution is relatively recent). In practice, advanced aliens might find it more efficient to send space probes to watch over worlds with the potential for civilization and report back (the basis of Arthur C. Clarke's *2001: A Space Odyssey*). Another search avenue that doesn't rely on ET taking a special interest in Earth is to look for signs of astro-engineering – advanced interplanetary building projects such as energy-harvesting 'Dyson spheres' that might give away their presence from their effects on starlight.

10 Talking to aliens

If we were ever capable of opening a communications channel with intelligent aliens, then how could we ever hope to exchange information with them? Figuring out potential communications protocols requires a huge effort of imagination to abandon all our human biases while identifying truly universal 'common ground'. The famous Arecibo message, broadcast to a distant, dense star cluster from the world's largest radio telescope in 1974, consisted of a series of 1679 binary digits (simple '1s' and '0s', the most basic form of counting there is). The number 1679 was chosen as it is the product of two prime numbers (another concept that should be universal to any civilization with a concept of number). Once reconstructed into 23 columns and 73 rows, the message produces a pictogram with basic information about the creatures that sent it.

TALK LIKE A GENIUS

❛ Some people think the requirements for "life as we know it" are needlessly strict, but they do have some good reasoning behind them. Carbon is by far the most common and versatile chemical element for making complex compounds; water's a widespread and stable solvent, for chemical building blocks to drift around in and eventually join up; and life itself is ultimately the organized harvesting of energy, so you're bound to need some kind of energy gradient it can take advantage of. ❜

❛ Of course, the panspermia theory raises the chances of widespread life – if complex biochemistry can get started on one planet and then get transferred somewhere else in a meteorite or on the back of a comet, then all bets are off. Life can probably find ways of hanging on in environments where it couldn't possibly have evolved in the first place. ❜

❛ Remember the famous Mars meteorite from back in the 1990s? NASA scientists thought they might have found traces of biological chemicals and maybe even fossilized microbes in a rock from Mars that got thrown into space and landed in Antarctica. Well, the claim's never really been disproved; geologists showed that some of the chemicals could possibly have been made without life, but that doesn't mean they necessarily *were*. We probably won't know for sure until we get geologists actually working on Mars itself. ❜

WERE YOU A GENIUS?

1 TRUE – the message does include a crude map showing Earth's location in the solar system.

2 TRUE – KIC 8462852 shows strange dips in brightness. An alien structure in orbit around the star is one plausible (though unlikely) explanation.

3 TRUE – searches traditionally focus on a few radio wavelengths that travel well across interstellar space, though many SETI scientists are now looking at a broader range of signals.

4 FALSE – the most famous unexplained signal was detected by the Big Ear radio telescope in 1977. Known as the 'Wow! signal' it never recurred but appeared to emerge from the constellation Sagittarius.

5 TRUE – this huge figure has been extrapolated from planet discoveries by NASA's Kepler satellite.

Suitable conditions for alien life are widespread in our galaxy, but that doesn't prove that life, let alone geniuses, are common.

Glossary

ADAPTATION
A physical or behavioural trait that aids an organism's survival and breeding in a specific environment.

ANALYTIC STATEMENT
In philosophy, a statement that can be shown to be true or false by analysing it without reference to other facts (*c.f.* synthetic statement).

ANTHROPOMORPHIC
An entity such as an animal that is nevertheless given human attributes.

A PRIORI / A POSTERIORI
An *a priori* proposition is one that is known to be true without evidence from experience. Propositions that can only be proved true from experience are *a posteriori*.

AXIOM
A statement that can be accepted without question in a particular mathematical system and used as a basis for developing logical proofs.

BLACK HOLE
A concentration of mass so dense that its gravity prevents light from escaping.

BOSON
Any subatomic particle with a whole-number value of the quantum property called 'spin'. Bosons include force-carrying particles such as photons.

CAPITAL
Financial wealth or physical assets that can be invested or deployed in order to generate a return through trade or other forms of business.

CAPITALISM
An economic system in which profit-making enterprises are privately owned and goods are traded in an open market subject to pressures of supply and demand.

CHROMOSOME
A carrier for genetic material found in the nucleus of cells of complex organisms.

COGNITION
The array of 'higher' mental abilities displayed by the human mind, such as reasoning, language, problem solving and memory.

COMPATIBILISM
A philosophical approach that attempts to reconcile the concept of free will with the apparently deterministic nature of the Universe.

COMPLEX NUMBER
Any number that is composed of both real and imaginary components.

CONSERVATISM
A political ideology that seeks to retain traditional institutions and values, often coupled with promotion of private enterprise and scepticism about government intervention.

CONSTANT
A number with specific value that plays a particular role in a mathematical or physical equation.

CONTINGENT
In philosophy, a fact that happens to be true, but in other circumstances might not have been (*c.f.* necessary).

DEMOCRACY
A form of government either formed from, or elected by, the entire eligible populace.

DETERMINISM
A view that every event is determined by the outcome of a prior cause. In philosophy this removes the possibility of free will. In mathematics and physics it implies predictability if the causal factors are precisely known.

DIALECTIC
The philosophical idea that any statement, action or state contains within it a contradiction that creates opposition, requiring the development of a synthesis that reconciles the two.

DNA
Deoxyribonucleic acid, a complex molecule that carries genetic information encoded in long sequences of chemical subunits called base pairs.

DUALISM
The philosophical view that the world is composed of two distinct elements, often perceived as the physical and mental realms.

EFFICIENCY
An economic term that describes both the distribution of resources in an optimal way (for instance to maximize the benefit to those who hold them), and techniques that maximize output while minimizing cost or other inputs.

ELECTROMAGNETIC RADIATION
A disturbance consisting of perpendicular electrical and magnetic waves that reinforce each other and propagate energy across space at the speed of light.

EMPIRICISM
The philosophical view that knowledge must be acquired through experience; hence there is no such thing as *a priori* knowledge.

ENLIGHTENMENT
An intellectual period spanning the 18th century, marked by great advances in philosophy and setting the stage for modern politics and economics.

EPISTEMOLOGY
The branch of philosophy concerned with the nature of knowledge, its limits, and techniques for establishing it.

EQUATION
A balanced mathematical relationship in which the terms on either side of an '=' sign are equivalent to each other.

FALLACY
An error of reasoning or false conclusion.

FALSIFIABILITY
The concept that any scientific theory should be capable of being proved false by new empirical evidence.

FERMION
Any subatomic particle with a half-integer spin. Fermions include all common matter particles.

FISCAL POLICY
The use of the overall government spending and taxation to influence the economy as a whole.

FITNESS
The degree to which an organism is adapted to its environment compared to others.

FRAME OF REFERENCE
Any fixed system of coordinates that can be used for measuring physical events. Einstein's theories of relativity show how measured properties can vary when two frames of reference are in relative motion or affected by gravitational fields.

FUNDAMENTAL FORCE
Any of four forces responsible for all known interactions in physics; gravitation, electromagnetism, and the strong and weak nuclear forces.

GALAXY
An independent system of stars, gas and other material with a size measured in thousands of light years.

GENE
The basic unit of inheritance. A gene is a subunit of DNA that produces a specific protein.

GENERAL RELATIVITY
Einstein's theory describing the nature of physics that occurs in the presence of large masses that distort four-dimensional spacetime.

GENOME
The full collection of genetic information within an individual of a particular species, including both its genes and its 'non-coding' DNA.

GLOBALIZATION
A system of free movement of goods, money and labour across international borders, linking to greater interdependence between nation states.

GRAVITATION
An attractive force created by objects with mass, which causes other objects to accelerate towards them. According to general relativity, gravitation arises from the distortion of spacetime around massive objects.

HEURISTIC
A 'rule of thumb' used by cognitive processes in the brain to rapidly (but not always correctly) assess information.

HISTORICISM
An approach to cultural practices and texts that considers them as the consequence of past events and current social context.

HUMANISM
A philosophical approach that considers humanity rather than the supernatural world as the centre of inquiry.

IDEALISM
A philosophical view that reality is ultimately immaterial, with physicality a mere projection or illusion created by the mental realm.

IMAGINARY NUMBER
Any number that is a multiple of the square root of -1 (denoted i). Although i cannot be calculated, giving it a symbol and treating it as a number in its own right leads to many useful mathematical results.

INFLATION
The annual rate at which the general level of prices in an economy is increasing.

INTEREST RATES
The cost of borrowing money, usually expressed as an annual rate.

KEYNESIANISM
An economic view that argues for managing economies by encouraging or reducing demand using both monetary and fiscal policy.

LIBERALISM
A political ideology based on the freedoms and rights of the individual citizen.

LIBERTARIANISM
A political philosophy advocating liberty and the exercise of free will, with little or no government interference or taxation.

LIGHT YEAR
A common unit of astronomical measurement, equivalent to the distance traveled by light in one year – 9.5 million million km (5.9 trillion miles).

LOGIC
A series of techniques for the construction of rational proofs based on simple axiomatic statements and various mathematical or philosophical methods.

MACROECONOMICS
The large-scale factors affecting an economy as a whole, including inflation, growth, interest rates and unemployment.

MARKET
Any environment in which buyers and sellers can exchange goods, services and money at agreed but variable prices.

MARXISM
A political and economic philosophy, derived from Karl Marx, that analyses society based on contrasts in power, money and social class.

MASS
A measure of the amount of matter contained with an object, linked to its inertia. Mass is thought to be created by matter interacting with Higgs bosons.

MATERIALISM
The view that reality is ultimately material, and that apparent mental or spiritual aspects arise ultimately from physical activity in our brains.

METAPHYSICS
The branch of philosophy concerned with the nature of reality.

MICROECONOMICS
Small-scale economic factors affecting individuals, households and businesses.

MODERNISM
A cultural outlook popular in the early 20th century founded in ideas of progress, technology, but also increasing atheism and doubt.

MONETARISM
The economic view that an economy is best managed by adjustments to the amount of money circulating within it.

MONETARY POLICY
The use of interest rates and other levers such as buying and selling of government debt to control the amount of money in an economy.

MONISM
The philosophical view that things are made up of a single element.

MUTATION
A random change to a gene on which selection pressures can act.

NARRATIVE
A sequence of events and incidents described in a work of fact or fiction.

NECESSARY
In philosophy, a fact that is true in any circumstances, and could not be otherwise (*c.f.* contingent).

NEURON
A specialized cell found in the brain and nervous system, with the ability to receive, process and transmit electrochemical signals.

NOUMENON
In philosophy, a 'thing-in-itself', that exists independently of our experience.

PHENOMENON
In philosophy, a 'thing as experienced', channelled through the filters of human consciousness.

PHENOTYPE
The outward expression of an organism's genome, including both physiological and behavioural traits.

PHOTON
A self-contained package of electromagnetic waves that allows light to pass through a vacuum and display particle-like behaviour.

POSTMODERNISM
A cultural view widespread from the late 20th century in which scepticism, irony and relativism are dominant, while attempts at 'grand narratives' are cast into doubt.

POWER
A number, written as the superscript to another, which indicates how many times the second number should be multiplied by itself.

PROTEIN
A complex molecule that acts as the building blocks for tissues within a living body.

QUANTUM PHYSICS
The field of physics concerned with very small scales on which subatomic particles exhibit wavelike properties such as uncertainty in energy and position.

RATIONALISM
A philosophical approach that sees knowledge of the world as something best acquired through reasoning rather than observation and experimentation.

REAL NUMBERS
Any number whose value could be used to represent a point on a continuous line stretching in both positive and negative directions from 0 to infinity (therefore distinct from imaginary numbers).

RELATIVISM
The idea that ethical and other judgements are dependent on their cultural and historical context, and even the language framework in which they are described, and therefore cannot be judged as right or wrong in any absolute sense.

RENAISSANCE
A cultural movement spanning the 14th to 16th centuries and marked by a revival of classical learning and a newfound spirit of humanist inquiry.

REPUBLIC
Any state governed by an elected parliament and head of state, in which the people are citizens rather than subjects of a monarch.

ROMANTICISM
A cultural movement of the late 18th and early 19th century characterized by a rejection of rationalism and a reaction to the Industrial Revolution.

SELECTION PRESSURE
An environmental factor that affects the chances of an individual reproducing and passing on its genes.

SEMIOTICS
Originating in linguistics, the study of the structure of language, and specifically the relationships between real-world phenomena and the 'signifiers' used to express them in communication.

SOCIAL CONTRACT
The implicit mutual cooperation between the mass of individuals and the government that rules them, often framed in terms of a negotiation of rights.

SOCIALISM
A political ideology based on shared ownership of means of production and the fair distribution of their profits.

SPACETIME
A four-dimensional structure composed of three space dimensions integrated with time in such a way that the dimensions can be twisted and 'traded off' against each other.

SPECIAL RELATIVITY
Einstein's theory describing the nature of physics that occurs during relative motion at speeds close to the speed of light.

SPECTROSCOPY
The analysis of precise wavelengths and colours of light emitted or absorbed by atoms and molecules in various physical situations. Patterns of dark or bright light can act as chemical fingerprints to indicate which elements are present.

STANDARD CANDLE
A star or other object whose intrinsic luminosity can be calculated by some independent means. Comparison between the object's luminosity and its *apparent* brightness therefore acts as a useful indicator of distance.

STANDARD MODEL
The widely accepted model of subatomic physics that categorizes elementary particles as either fermions (matter particles) or bosons (force-carrying particles).

STAR
A dense ball of gas that has collapsed into a spherical shape and become hot and dense enough at its centre to trigger nuclear fusion reactions that make it luminous.

STRUCTURALISM
A critical approach to texts and social practices that aims to understand them in terms of rules systems and language frameworks.

SYMMETRY
A property of mathematical objects (both geometric figures and other mathematical relationships such as laws of physics) that causes them to remain unchanged following a manipulation known as a transformation.

SYNTHETIC STATEMENT
In philosophy, a statement whose truth can only be determined by checking the facts it refers to (*c.f.* analytic statement).

UNCERTAINTY PRINCIPLE
A quantum mechanical law that prevents certain pairs of properties (for example, position and momentum) from being measured with absolute precision simultaneously.

UTILITARIANISM
An ethical stance that judges the morality of an action on its consequences, aiming to produce the greatest benefit for the greatest number and minimize harm.

VIRTUAL PARTICLE
Any particle that can be briefly created out of nothing thanks to the uncertainty principle linking time and energy.

Index

ABOUT THE AUTHOR

After studying astronomy and science communication at the University of London, Giles Sparrow embarked on a two-decade career in publishing that has seen him write books on subjects as diverse as spaceflight, archaeology and mythology, as well as editing bestsellers from leading authors on topics ranging from quantum physics and economics to evolution and philosophy.

A keen quizzer when he gets the time, Giles has appeared as a team captain on the BBC's *Only Connect* and is currently writing his first novel, a historical thriller set in Regency England. He lives in Wapping with his partner, two cats and far too many books.

ACKNOWLEDGEMENTS

Just to be clear, I'm not claiming to be a genius – merely someone who's read a lot and worked with a great many smart people during two decades at the sharp end of publishing. Hopefully some of it's rubbed off! Thanks for particular inspiration from names such as Jim Al-Khalili, JV Chamary, Dan Green, Tom Jackson, Niall Kishtainy, Gemma Lavender, Darren Naish, Paul Parsons, Jonathan Portes, Marcus Weeks, Tat Wood and many others over the years, should in no way be taken as an endorsement by any of them of what's ended up on these pages.

Thanks are also due to Wayne Davies at Quercus for coming up with the initial concept and keeping a steady hand on the tiller during some of my wilder flights of fancy, to Natasha Hodgson for overseeing the project in-house, to Anna Southgate for going the extra mile when deadlines crunched, and to m'colleagues Tim Brown and Kaleesha Williams for pulling everything together.

This book is dedicated to my parents, John and Judy Sparrow, for encouraging my inquiring mind from an early age, and above all to Katja Seibold, for putting up with where it led me.

PICTURE CREDITS

17: Shutterstock/Calmara; 39: Shutterstock/Andrey Oleynik; 55 Shutterstock/HuHu; 66: Shutterstock/okili77; 79: Shutterstock/Bygermina; 87: Shutterstock/Steve Collender; 99: Shutterstock/Eric Issele; 119: Shutterstock/Ronnie21; 166: Agarzago via Wikimedia; 167: Wolfgang Beyer via Wikimedia; 169: Shutterstock/Tribalium; 177: Victor de Schwanberg/Science Photo Library.

All other illustrations by Tim Brown

First published in Great Britain in 2017 by

Quercus Editions Ltd

Carmelite House
50 Victoria Embankment
London EC4Y 0DZ
An Hachette UK company

Copyright © Giles Sparrow

Packaged by Pikaia Imaging
Copy editor: Anna Southgate
Design assistant: Kaleesha Williams

The moral right of Giles Sparrow to be identified as the author of this work has been asserted in accordance with the Copyright, Designs and Patents Act, 1988.

All rights reserved. No part of this publication may be reproduced or transmitted in any form or by any means, electronic or mechanical, including photocopy, recording, or any information storage and retrieval system, without permission in writing from the publisher.

A CIP catalogue record for this book is available from the British Library

ISBN 978 1 78648 529 8

Every effort has been made to contact copyright holders. However, the publishers will be glad to rectify in future editions any inadvertent omissions brought to their attention.

Quercus Editions Ltd hereby exclude all liability to the extent permitted by law for any errors or omissions in this book and for any loss, damage or expense (whether direct or indirect) suffered by a third party relying on any information contained in this book.

10 9 8 7 6 5 4 3 2 1

Printed and bound in Great Britain by Clays Ltd, St Ives plc